IN MEMORIAM

WAYNE E. AMUNDSON
(FEBRUARY 1956–SEPTEMBER 2014)

WITHOUT YOUR SUPPORT,
THIS BOOK WOULD NEVER
HAVE BEEN WRITTEN.

DEDICATED TO JENNIFER BROWN, PHOTOGRAPHER EXTRAORDINAIRE.

First published in 2009 by Voyageur Press, an imprint of The Quarto Group, 100 Cummings Center Suite 265D, Beverly, MA 01915 USA. This edition published in 2019.
T (978) 282-9590 F (978) 283-2742 www.QuartoKnows.com

10 9 8 7 6 5 4 3 2 1

ISBN: 978-0-7603-6415-4

Digital edition published in 2019
eISBN: 978-0-7603-6416-1

The Library of Congress has cataloged the previous edition of this book as follows:

Library of Congress Control Number: 2018967569

Acquiring Editor: Dennis Pernu
Project Manager: Alyssa Bluhm
Art Director: Cindy Samargia Laun
Cover Designer: James Kegley
Layout: Danielle Smith-Boldt

Front cover: © *Shutterstock, AnastasiaPash (main), Ekaterina Bratova (top left), Shcherbakova Natalia (middle left), Artsiom Petrushenka (bottom left)*
Back cover: *Terrapin Acres*
Title page: *Barb O'Meehan*

Printed in China

CONTENTS

ACKNOWLEDGMENTS

I am fortunate to have the support of many people. Everyone who reads this book will be getting not just my knowledge, but that of the people who helped me along the way.

Many of the photos between these pages were taken by me as well as family friends. Some have been gleaned from other breeders. I could not have completed this project without the wonderful photography of my friend Jennifer Brown. Jen came to the farm to get a kitten. Her camera work with the animals during her visit was so inspired that I pleaded with her to take photos for the book. She even stuck with me through some of the more disturbing medical procedures!

My farm partner and right hand is David Weber. In 2006, as a friend of our family and animal farming partner, Dave registered his first goats under the herd name Cutter Farms. Dave started keeping his animals here at the farm and took over some of the heavy lifting. He helped write and revise this book and provided some photographs. Since before the death of my husband, Wayne, Dave has helped keep the animals on the farm healthy and happy.

Poplar Hill will always seem like home after several years in a cooperative effort with Vince and Christine Maefsky. I have met quite a few goat owners who started with one or two pet goats and now face a hobby gone wild. One of these breeders is Vince Maefsky. Poplar Hill Dairy Goats—the largest goat dairy in Minnesota and one of the largest in the country—is only a few miles from our farm. I met the Maefsky family through the Minnesota Dairy Goat Association. (I also co-owned a buck with their daughter, Sarah Johnson.) Vincent picked up milk at our farm twice a week during my first commercial venture. I learned much of my dairy knowledge through Vince's stories and travels as he put up his beautiful dairy barn. Later, after our cheese plant no longer purchased milk from producers and we moved to another location, I took Vince and Chris up on their offer to move my milkers to Poplar Hill. Milking five mornings a week from three to seven (the time it takes to milk some five hundred goats in a forty-head modern milking parlor) was an education worth a degree. Vince and Chris, with their experience raising goats for more than thirty-five years, should be the ones writing this book—except I suspect they are too busy!

My friend Anna Boll Johnson was the first child I guided into the world of goats. Anna started helping on my farm at the age of thirteen. She has since had little girls of her own who visit to pet the goats. My daughter came along while Anna was still a girl helping me out. Viveka provided me with additional experience by showing me animal husbandry through the eyes of a child. Her picture (at various ages) can be seen throughout this book. Viveka even got into her university, in part, because of her college essay about being "raised by goats." Kids—human and goat—go together naturally.

During my years raising and learning about goats, I have had the opportunity to get to know a lot of fascinating people. I can't even begin to thank those who gave me my first lesson in goat care. I am indebted to my friend Lynn Litterer, who took me to a class at the University of Minnesota about raising Angora goats—long before I moved to the country or had space for them. The Minnesota Dairy Goat Association, whose newsletter (edited at the time by Mark Boorsma) taught me some basics, led me to ads for breeders so I could buy my first animals, and then trapped me into the editor position for three and a half fascinating years.

I have fond memories of the breeders of my first animals. Maddie Frounfelter of Landmark Oaks taught me how to milk and sold me Celestial Star. Linda Libra provided my second animal, Marshland DD Menolly. Linda became a friend and mentor and also allowed her daughters to help with my herd. Gloria Splinter of Hidden Springs Game Farm shared her exotic livestock expertise about Pygmies and Nigerians and introduced me to my first Fainting goats when her husband, Roland, brought some home from a sale in Iowa. Gloria and her vet tech daughter, Tara, even listened and gave ideas as I read parts of this book aloud over the phone.

At goat meetings and shows, I met people who will always be available to me for advice and help. Lucresha Larson, "the midnight milker," has a full medicine chest and is willing to share—as long as I don't call much before noon. Bev and R. J. Nohr of Charis Manor and Deb and Sharla Macke of Raintree and Calico Acres Dairy Goats are breeders I can always call on when I need a shoulder to cry on or have a triumph to share.

I wish to extend thanks to Annette Maze of Hill Country Farms. She graciously allowed a drop-in visitor to have a complete tour of her ranch at the time of the first American Boer Goat Association Convention in Texas in 1994. Visiting a full-fledged embryo transplant facility and hearing her many stories about her years working with goats added another valuable layer of experience.

The internet has allowed me to expand my knowledge exponentially. With a click of a mouse, I can ask goat owners across the country for help with a sick animal. I highly recommend this resource, with the caveat to read a lot—and don't believe everything you find!

Some online resources need mention. Linda Campbell of Khimaira Farms has an excellent website, ComeToTheFarm.com, that is full of advice. Larry Bunton of Fias Co Farm designed my logo. His ex-wife, Molly Bunton, has since moved from Tennessee to Michigan with the Molly's Herbals natural goat products company. I am thrilled that Molly continues to keep up with the Fias Co Farm website, a wealth of goat care advice and experience. Much of her information is geared to natural medicine. While I have never spoken to Joyce Lazarro of Saanendoah, her information on copper deficiency in goats helped me tremendously, and I highly recommend her resources. More recently, I have found lively discussions and suggestions on several of the many goat-related Facebook groups. Big warning here: I have also read some truly appalling advice, some of it from so-called "veterinary groups." Please take online advice with a grain of salt. And don't rely on a single expert—choose the advice of multiple sources you trust.

Thanks are also due to Drs. Tim Johnson and Bruce Oscarson; Osceola Veterinary Services; Hill Country Farms; John White, One-Oak-Hill; and Dr. Cindy Wolff, University of Minnesota Small Ruminant Specialist.

I am sad to report that on September 15, 2014, I lost my husband, Wayne Amundson, to the ravages of diabetes and heart-related illness. While working for a publishing house, Wayne encouraged me to write the first edition of *How to Raise Goats*. He suffered through the writing, helped with rewrites, and dealt with the angst of a novice writer. When we first got into goats, Wayne wasn't used to animals. He had no idea where my critter fixation would lead. Yet together, we showed our first kids, built a milking facility, and managed an increasing herd. As we traveled the goat show circuit, he also made friends and even served as a show secretary for some local club shows. Rest in peace, Wayne. You are missed!

Whether in the show, farm, or barn, goats are expressive, active creatures. Raising goats can be an extremely rewarding experience.

INTRODUCTION

SHE KNEW HOW TO LOOK AFTER THE GOATS AS WELL AS ANYONE, AND LITTLE SWAN AND BEAR WOULD FOLLOW HER LIKE TWO FAITHFUL DOGS, AND GIVE A LOUD BLEAT OF PLEASURE WHEN THEY HEARD HER VOICE.

—JOHANNA SPYRI, *HEIDI*

It is hard to imagine that *How to Raise Goats* has been in print since 2009. This third edition includes updated statistics and references, new science, some new photos, and things I have learned since writing the original. I can tell you, 2018 has given me vindication—goats are amazing. Previous lore says goats are negative or bad, as in "the hero or the goat." This year, the Merriam-Webster dictionary added a new definition. *GOAT* is now defined as "the greatest of all time," a statement that certainly describes my feelings about my favorite animal.

As a city girl, I'm not sure where my love of goats began. It might have been the eager friendliness of petting-zoo goats or early exposure to the book *Heidi* by Johanna Spyri. The antics of Little Swan and Little Bear on the Swiss mountain have fascinated many children and adults over the years. I was also drawn to the comfortable size of goats compared to cows or horses. Certainly, my fascination predates the myriad memes and videos filling the internet with goats. Whatever the inspiration, when I moved to the country, my first hobby-farm animal was a goat.

Poplar Hill Dairy started out as a hobby—and in Vince Maefsky's words, "went wild"—first starting out in this old dairy barn and then expanding to a modern, world-class goat operation. *Jen Brown*

Working in a pit milking forty goats at a time was an education. *Jen Brown, Poplar Hill*

The reasons people keep goats are as widespread as the places goats are kept—and goat owners are as varied as the animals they keep. Sometimes called the "poor man's cow," a dairy goat can comfortably supply a family with milk while being kept on a fraction of the space and resources a cow requires. This makes the goat an ideal animal for the back-to-the-land movement that surges every few decades. Goats are therefore the animal of choice for home-school families living on a small acreage.

Goat projects teach children the responsibility of caring for a living creature. Easy to handle, gentle, and gregarious, goats make a great introduction to raising livestock. With an increase in number and availability of meat and miniature goats, as well as dairy, children have more choices than ever before to raise and exhibit these fascinating creatures. National FFA Organization and 4-H goat projects are popular in many parts of the country. Through those organizations and others, I have met some bright, caring young people while showing and breeding goats. A number of these kids previously had little or no experience with livestock.

Young people aren't the only ones who can benefit from keeping goats, though. As the US population becomes less rural, adults and children have less contact with the land and the animals raised in the country. Learning about the cycle of life by owning a farm animal is one of the most enriching experiences imaginable. Increasingly, immigrants in the US are raising their own goats for an opportunity to use goat products that were prevalent in their home cultures. They too may have known goats only because of eating the meat or drinking the milk without any hands-on experience.

The practical uses for goats and their products are many. I keep goats as a multipurpose animal, and try to use all the products my goats provide. I eat their meat as well as drinking their milk. I've experimented with hides and bones for different projects. Some of my friends won't touch goat meat because they feel that doing so would be like eating a pet. I must admit that some of my own goats never reach the freezer for this very reason!

This book is a compilation of my wanderings and an introduction to the caprine world, including the science behind caring for goats. If you have an interest in keeping goats, I hope to give you an idea of what to expect—and how to get help when your capricious friends don't do what you expect. There are as many ways to keep and care for goats as there are goat keepers. Enjoy your journey!

Terrapin Acres is the name of my goat farm in Scandia, Minnesota. Our first goats were Nubian dairy goats. *Terrapin Acres*

1 A GOAT IS JUST A GOAT—RIGHT?

PEOPLE ARE GOATS, THEY JUST DON'T KNOW IT.

—CHES MCCARTNEY, THE GOAT MAN

When I decided to raise goats, there were just a few recognized dairy breeds. The LaManchas, with their funny little ears, were easy for me to rule out—they looked too goofy. The Swiss breeds were nice, just not what I wanted. Since I wanted a dairy goat, I didn't consider the tiny Pygmy or the silky-coated Angora. The heavily muscled Boer goat hadn't reached this country yet, the Nigerian Dwarf goat wasn't really considered a milker, and the stiff-legged Tennessee Fainting goat wasn't even on my radar. The Nubian, with her long ears and variety of colors, drew my eye. Just like that, my choice of goat breed was decided, and a journey I'd never anticipated began!

I'm far from alone in my fascination with goats. Geographically, goats are the most widespread livestock species. Because of their adaptability, goat herds can be found anywhere from the cold mountains of Siberia to the deserts of Africa to the moist regions of the tropics.

Scientists think that the wild bezoar (*Capra aegagrus*), an ancestor to the goat that still thrives in Europe and Asia, was the first domesticated herbivore. In the Fertile Crescent of the Middle East, archaeological studies have found evidence that goats were bred in early human settlements. Goat remains excavated from a cave show signs of a healed broken leg that could only have been treated by humans. Those remains were carbon-dated to between twelve and fifteen thousand years ago.

Other studies suggest that Asia was another center of caprine development. Through DNA analysis, researchers have found a completely different strain of Asian goats, along with unique strains of pigs, cows, and sheep. Wild goats in Pakistan may have been the ancestors of Cashmere goats. Many traits of the Cashmere are different from the rest of the goat breeds.

Despite these multiple origins, animal geneticists have concluded that the world goat population has less genetic variation than the world cattle population. Goat DNA shows only about a 10 percent variation between animals on different continents. Cattle DNA varies by 50 percent or more. This lack of variation seems to indicate that goats migrated with their keepers across continents.

Migratory people likely took along the most versatile animals. Goats provided milk, meat, fiber, and hides while also carrying some of the load on their backs or pulling carts of gear. Their adaptability and their calm, friendly nature make goats easy to transport. In colonial times, they traveled on the Mayflower and other ships to provide food. These goats stayed with colonists or were sold as trade goods. Some were released on islands to run wild to provide supplies for sailors on future voyages—not a great idea. In the absence of natural predators, the nonnative goats soon overran their island homes. The ecological damage from released goats was devastating. These browsers efficiently stripped vegetation. In the Galapagos Islands, goats were a leading cause in the demise of the famous giant tortoises.

Interest in goats expanded rapidly through the 2000s, partially due to the growth of ethnic groups in the United States, as well as incentives in the southeastern states for tobacco farmers to switch to other sources of income. Since I first wrote this book, the number of goats inventoried by the United States Department of Agriculture (USDA) has dropped by one million animals. The January 1, 2018, inventory by the USDA totaled almost 2.62 million head. This number included 2.16 million breeding goats. I can't be certain if the dropping census numbers are also due to actual decreases in the number of goats or if they reflect decreased reporting and fewer commercial farming enterprises.

Opposite: Goat horn has traditionally been used for musical instruments, such as the Jewish shofar. *Shutterstock*

The range of color and form found in goats is amazing! *Jen Brown, Terrapin Acres*

The numbers include 380,000 dairy goats, 142,000 Angora goats, and 2.1 million meat goats. Dairy goats represent 15 percent of all goats in the United States. Meat animals account for 80 percent of goats, while Angora goats number the remaining 5 percent. It is rare to find an Angora herd of any size outside of Arizona, New Mexico, and Texas. In fact, before 2005, the USDA only surveyed three states. The survey method and status of goats as minor species in the agricultural community most likely vastly underestimates the true number of goats.

In reality, it is impossible to guess how many goats live in the United States. The agriculture census is not a survey of pets or hobby farm animals. With increasing numbers of ethnic communities, the demand for traditional foods from their native lands continues to grow. Others are simply interested in food from sustainable sources. Miniature goat breeds are popular pets. Only a cursory search on the internet brings up a dizzying number of goat sites, videos, and photos. It would seem that interest in my caprine friends has not waned since they became human companions at the dawn of civilization.

The use of goats and their products is limited mostly by public perception. Some societies prize goats; others view goat products with suspicion. Goats make excellent companions and beasts of burden but are better known for the products they provide. These products vary as widely as the animals themselves. Obvious commodities include cheese, milk, hair, and meat. But goat hide, a byproduct of the meat-goat industry, has many uses. Goatskin parchment was once used for writing in Europe. Wine and water were carried in containers made from goat hide. Glue has been made using hide as well as hoof and horn shavings. A soft, fine goatskin called morocco is imported from Africa today and used by modern bookbinders because it is strong and durable. Musical instruments, especially drums, are made from goat hide. Goat leather also makes fine boots and clothing. The phrase "kid gloves" still brings to mind the finest of soft leather. In fact, my favorite gardening gloves are made from goatskin! Modern "catgut" used as a medical suture is often made from goat intestines. Elite car manufacturers use Boer goat leather due to its beauty and durability. Even the horns of goats have numerous uses.

The modern flutophone probably originated from the gemshorn (German for "goat horn"), a medieval instrument played like a flute and made from goat horn. The shofar, a ceremonial horn used in Jewish tradition, is sometimes made of goat horn. Goat horn has been used to make eating utensils and drinking containers. Goat horn, as well as bone, is still used today for jewelry, buttons, and crafts.

GOAT ASSOCIATIONS AND SOCIETIES

The first two editions of this book contained addendum pages listing various clubs and registrations for goats. With the growth of the internet and ubiquitous access to computers, I have removed this section in favor of recommending new goat owners do their research online.

A few people get into goats by chance. They see some cute goats and just buy them at auction or from an animal swap without any planning. More often, there is a certain amount of research that starts with the question, "What kind of goat should I get?" Various goat groups and registry associations can answer that question and many others. These organizations are wonderful resources for goat owners to learn about the breeds, get to know other goat enthusiasts, and participate in sponsored shows or events. Some groups are simply clubs of enthusiasts or local branches of national registries. Often, your choice will be made easily by finding out which group is most active in your area.

I got my first information about goats by joining the Minnesota Dairy Goat Association. This local group is affiliated with the American Dairy Goat Association (ADGA) and sponsors shows, youth trainings, and educational activities for dairy goat owners in my state. You can find your own local clubs by an internet search or by contacting the national organization about what clubs are operating in your area.

The American Milch Goat Record Association began registering dairy goats in 1904 and published the first herd book in 1914. (Milch derives from an Old English word meaning "giving milk.") Now known as the American Dairy Goat Association, this venerable organization has almost 15,000 members.

The ADGA recognizes seven full-sized and one miniature breed as dairy goats in their registry. Efforts are underway to add the Golden Guernsey goat to this list under the name Guernsey goat. As American breeders work with the genetics of a few imported embryos and semen from the British Isles, this rare breed may find new life in a new land.

The American Goat Society (AGS) is about one-tenth the size of the ADGA. AGS was founded in 1935 as a registry organization for only purebred goats. Today the AGS registers purebreds from the eight dairy breeds accepted by ADGA as well as a purebred Pygmy goat herd book.

In ADGA, when two different breeds of registered "Purebred" goats mate, their offspring are eligible to be recorded as Grade or Experimental. Purebreds that have serious enough defects to disqualify them from being registered in their breed must also be recorded as Experimental. For example, Sable Saanens were listed as Experimental before being recognized as a separate breed from traditional white Saanens.

ADGA permits the registration of goats designated by breed name with the addition of "American." A Purebred dairy goat comes from a Purebred sire and Purebred dam of the same breed, conforming to breed standards. An American goat is the offspring of a sire and dam of the same breed going back a minimum three generations for does and four for bucks. ADGA maintains separate herd—official lists of registered animals—for Purebred and American goats. The LaMancha and Sable breeds have an open herd book, meaning that goats can "breed up" into the Purebred registry. Nigerian Dwarf goats have only a Purebred herd book with no Grade program. Currently, crosses of Nigerian Dwarf goats with full-sized dairy breeds are not recognized or registered by the dairy goat association, although there are registries and clubs for these "minis." AGS is considering the possibility of adding American herd books into their registry.

The International Dairy Goat Registry (IDGR) was created in 1980 to provide low-cost registration and record-keeping service for owners of all breeds of goats and sheep. With herd books for Purebreds, Americans, Grades, and Experimentals in every breed, IDGR tries to support every goat and sheep breed. By October 2014 they changed their name to the International Dairy Goat Registry-International Fiber Breed Registry to better reflect their mission. Most recently, they became the International Goat, Sheep, Camelid Registry. From the looks of things, they register all manner of dairy goats, including "mini" versions of every recognized dairy breed crossed with

Nigerian Dwarf goats. Theoretically, they will also register meat goats, but none are currently listed on their website.

In early 2000, the ADGA and the American Meat Goat Association (AMGA) began a joint effort to create an all-inclusive national goat organization. This group is trying to represent the multiple goat-related industries, and by 2009, the American Goat Federation was formed. This group provides support and education for the dairy-meat, and fiber-goat industries. It does not offer animal registrations.

TYPES AND BREEDS OF GOATS

Despite the genetic similarities among the goat population as a whole, experts have identified more than three hundred distinct caprine breeds. A breed is defined as a group of animals that has certain traits in common, such as similar color, conformation, function, or size. Animals within a breed pass these traits to their offspring. Landrace breeds are those that evolved in the wild through natural selection and are ideally suited to their specific environment, while modern domestic breeds are created through controlled matings. These modern breeds of goats are often separated into their respective breeds by main purpose—easily researched by following information put out by the societies and associations formed to promote them.

DAIRY GOATS

Dairy goats are distinguished by ear type, color pattern, and size. The Swiss breeds (Alpine, Oberhasli, Saanen, Sable, and Toggenburg) have upright ears. These breeds are distinguished from one another by color.

Alpine

The ancestors of all purebred Alpines in the United States today arrived in 1922,

ALPINE COAT PATTERN

Alpine coat patterns are expressed in French terms. When a recognized coat pattern is banded or splashed with white or another color, it is described as broken, as in "broken chamoisee."

- **chamoisee.** Brown or bay markings with a black face, dorsal stripe, feet, legs, and sometimes a martingale running over the withers and down to the chest. A two-tone chamoisee goat has light front quarters with brown or gray (not black) hindquarters. (The male version is chamoise.) Broken chamoise has a band or splash of color "breaking" a solid color.
- **cou blanc** ("white neck"). White front quarters and black hindquarters with black on the head.
- **cou clair** ("clear neck"). Tan front quarters and black hindquarters.
- **cou noir** ("black neck"). Black front quarters and white hindquarters.
- **pied.** Spotted or mottled.
- **Sundgau.** Black with white markings on underbody and face.

Census figures from January 1, 2018, show 380,000 dairy goats in the United States. *Jen Brown, Poplar Hill*

The ideal Alpine has a straight face; a Roman nose is faulted. *Jen Brown*

when twenty-one animals were imported from France. Alpine goats are sometimes called French Alpines. The so-called Swiss breeds of dairy goat are thought by some to be color variants of the Alpine breed of goats—thus giving us Saanens, Sables, Toggenburgs, and Oberhasli to grow into separate breeds.

Eleven percent of the goats registered or transferred by ADGA in 2017 were Alpines. Prior to the addition of the Nigerian goat, Alpines were the second most popular dairy goat in the United States. The Alpine has medium to short hair and upright ears. The breed standard requires a straight face; a Roman nose (convex muzzle) is faulted. Toggenburg coloring (brown body and white markings) or all-white coloring is discriminated against. Mature does should stand 30 inches tall at the withers and weigh 135 pounds. Mature bucks should stand 32 inches tall at the withers and weigh 160 pounds.

Alpines are strong milkers and can often be found in commercial goat dairies. In 2017, the average milk production for Alpine does in a 305-day lactation was 2,611 pounds milk with 85 pounds fat and 75 pounds protein or about 304 gallons of milk. This stands to reason, since they are often reported to milk "a gallon a day." A gallon of milk weighs 8.6 pounds.

LaMancha

The name LaMancha originated from an unreadable description of some short-eared goats sent for exhibition to the 1904 Paris World's Fair. While the name was illegible, the words *La Mancha, Cordoba, Spain* were readable. This breed was probably originally known as Murciana.

Arriving in California with Spanish missionaries, these short-eared, dual-purpose milk-and-meat goats spread through the American West. In the 1920s, Phoebe Wilhelm crossed about 125 descendants of the mission goats with Toggenburg bucks. Later, Alpines, Nubians, and some Saanen bucks were bred to the short-eared does. Eula Fay Frey of Oregon worked hard to get LaManchas accepted for registry. In 1958, ADGA registered the first LaMancha—Fay's Ernie, L-1. About two hundred animals, sixty of which belonged to Frey's herd, started the registered herd book. The small ears are a distinctive breed characteristic and a dominant trait. Short ears carry even when crossed with another breed.

I didn't consider LaManchas for my first goats because of their appearance. This common initial reaction to the pixie-eared goat is quickly overcome by personal experience. The gentle, curious personality of the LaMancha wins people's hearts. In fact, in 2017, LaManchas made up 10 percent of ADGA registrations.

The LaMancha is no slouch in the dairy. In 2017, average production for a three-and-a-half-year-old LaMancha dairy goat on official Dairy Herd Improvement Association (DHIA) test was 2,245 pounds of milk, 84 pounds of fat, and 70 pounds of protein.

The "gopher ear" style on LaManchas measures less than 1 inch with little or no cartilage. This is the only type of ear eligible for registration of LaMancha bucks. *Jen Brown*

A dependable dairy goat, LaManchas milk for a long, steady lactation, giving high volume, butterfat, and protein. They are easy keepers. *Jen Brown*

In 1960, so-called elf-ears were prohibited on registered LaMancha bucks. The "elf ear" must be less than 2 inches long with the end of the ear turned up or down. Some cartilage shapes this small ear. *Jen Brown*

Nigerian Dwarf

Zoos initially brought miniature goats to the United States to feed large cats. The gentle nature of minis led to their popularity as pets. Nigerian Dwarfs and the Pygmy breed share the same genetic base, but over time breeder selection split them into two distinct breeds. The Nigerian Dwarf wasn't as common as the Pygmy. At one time, Nigerians were considered a "rare breed." The American Livestock Breeds Conservancy (ALBC) had Nigerians classified as "Recovering" on their Conservation Priority List.

The head, limbs, and body of a Nigerian goat are proportionate, a condition known as pituitary dwarfism. In 1981, the AGS was the first registry to recognize the Nigerian as a dairy goat. The IDGR started recording the Nigerians in 1982. The breed was accepted into the ADGA registry in 2005. Before these dates, Nigerian goats were considered solely a pet breed. Since ADGA recognition, the breed has flourished. Not only are Nigerians no longer listed as a conservation concern, the Nigerian Dwarf in 2017 was the most popular goat registered by ADGA by a long margin. The breed made up 35 percent of the registrations handled by ADGA with the next runner up, the Nubian, accounting for 26 percent.

The Nigerian Dwarf Goat Association registers Nigerian Dwarf goat does, bucks, and wethers. Their breed standard calls for slightly smaller goats than ADGA. NDGA also has a Milk Record Program. Unlike the ADGA DHIA, NDGA specifically requires does on test be no taller than 21 inches at the withers. Club shows include wethers, which are not shown or registered by the American Dairy Goat Association.

Kathleen Clapps of Texas was the first breeder to enter Nigerian Dwarf goats into an official milk test. Her goat earned Advanced Registry Star Milker status with 427 pounds of milk, 25 pounds of fat, and 20 pounds of protein. This level of production, not out of the ordinary for a traditional dairy goat, showed that Nigerians could successfully meet the standards set by the larger breeds. By 2017, the average 3-year-old Nigerian Dwarf doe on official milk test produced 778 pounds of milk, 48 pounds of fat and 34 pounds of protein.

One-third of a gallon of milk a day isn't bad at all for such a small animal!

Nigerian Dwarf goats enjoy people and can become attached to their owner. Of course, these traits may lead to a "talkative" goat that lets you know when she wants something! Because of their small size, Nigerians are a favorite with FFA, 4-H, and homestead families.

The original Nigerian-type goats from Africa were black, a recessive color. Crossbreeding established the breed and gave today's Nigerian Dwarf a variety of colors and patterns. Nigerian goats are one of the few breeds to have some individuals with blue eyes. The Nigerian breed standard set by the ADGA includes short, fine hair; a straight or dished face; and erect, alert, medium-length ears. According to ADGA standard, the does stand no more than 22.5 inches and the bucks no more than 23.5 inches at the withers. This differs from the Nigerian Dwarf Goat Association, which requires smaller heights of 17–19 inches for does and 19–21 inches for bucks with an absolute maximum of 23 inches at the withers.

Nubian

The Nubian is a combination of English goats and goats from other parts of the world. English ships traveled to many parts of the world, carrying goats to provide fresh milk and meat. Goats from these ports were crossed with common English milking goats.

The Nigerian Dwarf is a miniature breed of dairy goat. ADGA standards require does stand no more than 22.5 inches (57 cm) and bucks no more than 23.5 inches (60 cm). This height was established after years of wrangling by breeders. *Jen Brown, Cutter Farms*

During the 1800s, some goats were brought to France from Nubia, a region in North Africa. Descendants of these goats, called Nubians, arrived in England in 1883. Recognition for this long-eared breed came quickly. By 1896, the Nubian was a registered breed in England.

Nubian-type animals came into America as early as 1896; however, those bloodlines have disappeared. The beginnings of the first registered Nubians in America are traced to animals imported by J. R. Gregg of California. By 1918, forty animals were registered as purebred Nubians in the United States. It was a quiet beginning for the breed that has become the most popular dairy goat in the United States. Nubians previously outnumbered all other currently recognized dairy breeds two to one, before being dethroned by the Nigerian Dwarf in popularity. The stately Nubian still made up 26 percent of ADGA's 2017 registrations and transfers.

The Nubian is a dual-purpose goat, useful for both meat and milk. Known as the Jersey cow of the goat community, it produces a moderate amount of very rich milk. Averaging less milk than the Swiss breeds, Nubian milk has high average butterfat content, between 4 and 6 percent. The average Nubian lactation on official test produced 1,984 pounds of milk, 92 pounds of fat and 75 pounds of protein in 2017. For those purchasing a Nubian doe and being disappointed that she isn't producing "gallon a day," remember that the average 305 lactation has a peak production and then it may taper off. While a Nubian produces a gallon at peak, the production is a more realistic average of ¾ gallon with rich 4.7 percent butterfat daily.

With her long and graceful ears, a proud Roman nose, and a variety of colors, the Nubian is the most popular breed of dairy goat in the United States. *Jen Brown, Terrapin Acres*

The Nubian breeding season is longer than that of the Swiss breeds, so it is easier to breed Nubians for off-season milk. According to the breed standard, a mature Nubian doe should stand 30 inches at the withers and weigh 135 pounds. A male should stand 35 inches at the withers and weigh 175 pounds. Nubian hair is short, fine, and glossy. Any colors or patterns are acceptable.

Oberhasli

Beginning in the 1930s, an Alpine-type goat was imported into the United States from the Oberhasli region of Switzerland. The goat was originally known as the Swiss Alpine and considered a color variety of the Alpine breed. In 1979, the ADGA recognized the Oberhasli as a separate breed. The Oberhasli Herd Book began in 1980. In 2017, the average production of Oberhasli does in milk was 2,143 pounds of milk with 77 pounds of fat and 64 pounds of protein—or about 7 pounds of milk per day.

Like the Nigerian Dwarf, the Oberhasli was once listed as endangered by the ALBC. Today, more breeders are raising this attractive animal, which is now increasing in number and designated as recovering by the ALBC. Three percent of papers initiated or transferred in 2017 with ADGA were Obers.

Saanen

Saanens get their name from the Saanen Valley in the southwest of Switzerland. Starting in 1904 and until about 1922, about 150 Saanens were imported into the United States through Canada. These, in addition to later imported animals from England, formed the foundation of purebred Saanens in the United States.

Their white hair and pink skin has been said to make them prone to sunburn, cancer, and problems with heat in warmer climates. In

Mid-sized, vigorous, and alert, Oberhaslis are described as chamoise in color. Chamoise is a rich red color, preferred over lighter tones or black. Bucks often have more black on the head than does, plus more white hairs. *Susanna Yoemans, Cardinal Hill*

reality, with proper shade protection, Saanens thrive in many different climates. They are the most widely distributed breed of dairy goat in the world. In the United States, 6 percent of the dairy goats registered with ADGA in 2017 were Saanens.

Majestic white Saanens are a popular dairy animal because of their calm, eager-to-please temperament. Known as the Holstein cow of the goat world, Saanens produce, on average, the most milk of any of the dairy breeds. Average Saanens on test in 2017 produced 2,731 pounds of milk with 89 pounds of fat and 79 pounds of protein in 305 days. They are commonly known to give more than a gallon a day.

Sable

Due to a recessive gene in the original white Saanens, colored goats have always been a part of the Saanen bloodlines. It was speculated that Saanens developed in their isolated valley from white-colored Alpines preferred by local farmers. If this was the case, one would expect the occasional dark-colored offspring. The first Saanens were registered regardless of their color. In the late 1930s, the Saanen Breeders Association adopted a resolution that future Saanens had to be white to be acceptable for registry. Periodically, during following decades, Saanen breeders discussed the issue of Saanen kids that were perfect in every way except their color.

A letter published in *The Goat World* in September 1918 tells of a black kid born to two white parents. There was no existing Saanen or color restriction, so she was registered Saanen, like her sire and dam. But people kept asking the owner how it was possible that the kid was black. He answered when he registered her: "Damif-Ino."

By 2005 (as ADGA responded to members requests), the Sable was accepted as an official dairy goat breed for registry. By 2017, 4,640 Sables, or 1 percent of total goats, had their papers transferred or registered with ADGA. Being essentially a colored Saanen, DHIA records of a 305-day lactation show similar averages to the Saanens on test. In 2017, the average milk was 2,513 pounds, with 88 pounds of fat and 73 pounds of protein.

The Saanen is the largest of the dairy breeds, with rugged bones and plenty of vigor. White color is preferred, although light cream is acceptable. *Jen Brown, Poplar Hill*

A white pattern of markings is specific to Toggenburgs, including erect white ears carried forward with a dark spot in the center and white stripes running from above each eye down to the muzzle. *Jen Brown*

The Sable Herd Book is an open herd book like the LaMancha's. After three generations of American Sables in the pedigree, the next generation that meets breed standard is eligible for Purebred Sable status. A first-generation American Sable may have two Purebred Saanens as parents but meet the Sable requirements strictly based on color.

The reverse is not true. If two Purebred Sables produce an all-white offspring, that kid is now only eligible to be recorded as Experimental since the color standard for Sable specifies anything "except solid white or solid cream." In order for color to show in the offspring, both the sire and the dam have to pass the color gene to the kid.

Toggenburg

The Toggenburg is the aristocrat of the dairy goat community. Among the first purebred dairy goats to come to the United States, these animals had impressive family trees back home in the Toggenburg Valley of Switzerland. Swiss exporters in the early 1900s proudly claimed the breed had been pure for three hundred years. The Swiss Toggenburg breeders association calls the Togg "the oldest and purest breed in Switzerland."

At the turn of the previous century, these chocolate-colored goats with striking white trim were the most common dairy goat found in the United States. A famous herd of Toggenburgs was raised in North Carolina by Lillian Sandburg. (Her husband, Carl, was famous for his poetry, while she was well known in dairy-goat circles.) A case can be made that the modern dairy-goat industry was founded on the Toggenburg. This may come as a surprise to modern dairy-goat lovers, since in 2017, only 3 percent of the registration papers either granted or transferred by ADGA were Toggs.

Sometimes called deerlike in stance and appearance, the Toggenburg has longer hair than the other dairy breeds. The coat can be shorter in America because no emphasis has been placed on hair length. In Switzerland and England, long hair is desirable on the shoulders and back legs.

Size is another breed characteristic that changed as the animal moved from its native country. Swiss and British purebred Toggs are small, economical animals. Meanwhile, in the United States, Toggenburgs have increased in size. A few years ago, ADGA standards changed from 125 pounds and up to 115 to 150 pounds.

The breed has won many best udder awards. According to the DHIA, the all-time milk producer for all breeds is a Togg doe named SCH Western Acres Zephyr Rosemary who produced 7,965 pounds of milk in 305 days in 1997. Many Toggs milk into their early teens. Of course, the breed average is lower than Rosemary's (yet still respectable): 2,254 pounds of milk, 70 pounds of fat, and 63 pounds of protein.

FIBER GOATS

The most common fiber goat in the United States is the Angora, with 142,000 tallied by the NASS All-Goat Survey in January 2018. The hair of the Angora goat is made into mohair. The United States is one of the largest suppliers of mohair in the world. Cashmere goats are less numerous and produce the fiber known as cashmere. Crosses are popular with hobbyists and hand-spinners.

Registries exist for Angoras, Cashmeres, and Colored Angoras. The oldest registry for purebred Angora goats is the American Angora Goat Breeders Association, formed in 1900 to keep records of purebred animals. The Cashmere goat Association was founded in 1992 (originally the Eastern Cashmere Goat Association) to promote Cashmere goats and their products. And as mentioned previously, the International Goat, Sheep, and Camelid Registry registers all types of goats.

The fiber-goat industry has been in existence in the United States for more than one hundred years. Most of the industry's commercial herds are in Texas, with 90 percent of these Angora goats kept primarily on range. Some of the most luxurious cloth in the world comes from goats, and interest in natural fibers worldwide is increasing. Unfortunately, the industry in the United States has struggled due to climate and pricing issues. In better news, reports in spring of 2018 indicate that mohair fleece prices are up by 30 percent.

Many of the fiber goats are owned by small herds catering to hand-spinners and crafters. These goats vary from the commercial Angora, which has a standardized white coat and look. The owner of small fiber goats may be interested in colored goats, smaller goats, or the dual-purpose fiber-meat goat, the Cashmere. Whatever the type, these goats are most often kept for the beautiful fleeces.

Angora

Angora goats originated in Asia Minor. In Turkey they were a prized commodity for two thousand years. These plush goats came to the United States in 1849. Dr. James Davis of South Carolina helped introduce cotton production to Turkey. The Sultan was so pleased with the new plant fiber that he sent Dr. Davis a gift of seven Angora does and two Angora bucks. Purebred Angora goats had just arrived eleven years earlier in South Africa. By the late 1800s, Turkey banned the export of further Angora goats. However, the earlier introductions allowed both South Africa and the United States to dominate modern mohair production.

The Angora is a smaller breed of goat than most dairy goats and tends to have fewer kids. The breed is quiet and friendly. Some descriptions suggest they are shy to the point of being timid. Both sexes have horns, the buck having long horns with a pronounced spiral. The doe's horns are shorter and usually grow straight back with little spiral. Hair being the predominant feature of the Angora, this animal has hair from the eyes down. Mohair can grow 10 inches or more in a year, hanging in tight ringlets on the body from belly to back and along the sides and extending along the neck and head. The tighter the twist in these ringlets and the denser the growth, the higher

The Angora goat is popular across the country with hand-spinners and hobby breeders. Breed standard for purebred Angoras is white. *Shutterstock*

quality of fleece and more valuable the goat. This fleece is normally sheared twice a year, with the average fleece weighing about 5.3 pounds. There are some Angoras with blue eyes.

Cashmere

Adult Cashmere goats are sheared or combed once a year and yield as much as 2.5 pounds of fleece. The hair on Cashmere goats comes in two types: harsh guard hairs and fine underwool. The fleece must be dehaired to remove the guard hairs from the luxury fibers.

The Cashmere goat is not a breed, but a type of goat. Cashmere wool is grown by all caprine species except the Angora. A Cashmere goat is one that produces underwool of commercially acceptable color and length. Cashmere goats are judged by the quality and quantity of the underwool and the size or build of the animal. The quantity and quality of cashmere fiber produced are determined by the crimp (or style), the diameter and length of the fiber, and how much of the animal's body is covered with downy underwool. After the soft, downy fibers are cleaned from the guard hairs, the average 2.5-pound fleece produces only 0.25 to 0.5 pound of cashmere. The effort is worth the trouble as cashmere is considered the lightest weight, warmest, and least irritating animal fiber.

The Indian state of Kashmir exported prized shawls to England that were fine enough to pass through a wedding ring; hence, they were called "ring shawls." One might think that since the name Cashmere comes from Kashmir and the surrounding mountains of India and Pakistan, that the goats are also from that region. In reality, the fiber came from goats in Tibet to be woven by artisans in Kashmir. As an industry in the United States, cashmere production is very recent on any commercial scale. In the late 1980s, Cashmere goats were imported to the United States from Australia and New Zealand to be selectively bred with Spanish goats and others in the United States for improvement of the herds. The North American Cashmere goat is now being registered.

The Cashmere Goat Association exists to promote these animals, as well as provide breeder education. The Cashmere Goat Registry allows animals meeting standards to be registered with them. The standards do not discriminate by size or color, although spotted goats are discouraged. Main judging criteria involve the production of cashmere as well

A Cashmere goat produces soft, high-quality fiber, prized for sweaters and other garments. *Shutterstock*

as body. Registrants are required to submit a fleece to meet specific standards. Fleece must not exceed a required Average Fiber Diameter (AFD), a characteristic that increases by age, so this number is age dependent. Yearling fleece cannot have an AFD greater than 14.5 microns, while the AFD on a nine year-old goat can be as high as 18.5 microns. The International Goat, Sheep, and Camelid Registry and Pedigree International also registers Cashmere goats.

Cashmeres are generally raised as dual-purpose animals for fiber and meat or as brush control. The goats have been bred to have wide horns, blocky builds, and refined features. Because their feral origins are more recent than those of other breeds, Cashmere goats tend to be wary rather than placid. Breeders report that they are easy kidders and good mothers.

A cross between an Angora and a Cashmere goat is called a Cashgora. Its coat has more of a crimp than cashmere and fewer guard hairs. The fiber is said to combine cashmere softness with Angora curl and to take dye nicely.

Colored Angora Goat

Until recently, Angora goats were bred solely for white mohair. Angora goats born a color other than white stood out in commercial flocks where standardized color is important. As a result, those animals that didn't meet the color standard were being sold for meat or destroyed. Small holders interested in hand-spinning a different color mohair began rescuing and breeding these castoffs. Thus, the Colored Angora goat became a separate recognized breed.

By 1999, the Colored Angora Goat Breeders Association was formed. In 2002, breeders established the American Colored Angora Goat Registry in order to register and record the bloodlines of purebred Colored Angora Goats. These goats are now seen with fleeces of black, gray, silver, red, and brown in both solid and patterns. The proper Colored Angora goat should have color expressed throughout the fleece. The patterned goat fleece should contain no more than 60 percent of fibers. As interest in these animals increases, more clubs and registries are forming. Good resources also include spinning and fiber clubs.

The Pygora is essentially a Pygmy Angora goat, possessing the fine-hair coat of the Angora and the small stature of the Pygmy. *Fran Bishop, Rainbow Spring Acres, Pygora Breeders Association*

Nigora

Crossing small-breed goats with their large-breed counterparts has become popular. A cross between the Nigerian Dwarf and the Angora, the Nigora is a miniature fiber breed. These goats come in a variety of colors. They may have mohair or cashmere fiber, but the typical fleece falls into the category of cashgora. The American Nigora Goat Breeders Association was formed in 2007.

Pygora

The Pygora is another small-breed cross. Angora does were bred to Pygmy bucks. The short, fuzzy Pygoras are popular with hand-spinners, hobbyist breeders, and pet owners.

The Pygora Breeders Association, formed in 1987, has seen an increase in the popularity of its breed. Breed standards require that the registered Pygora goat be no more than 75 percent registered Angora goat or 75 percent registered Pygmy goat. The offspring of an Angora and a Pygmy is considered not a Pygora but simply a cross; these animals are registered as first generation.

The Boer goat is a popular meat breed. *Jen Brown, Cutter Farms*

MEAT GOATS

Meat goats are the most numerous type of goat in the United States, with 2.12 million head at the start of 2018. The census in 2007 showed approximately 3 million head. Through 2009, meat-goat numbers grew at 3 to 5 percent each year and made meat goats the most rapidly expanding type of goat in the United States. Downturns in the economy and recent years of drought in the Southwest, notably in Texas, have reduced numbers of meat goats since that time. Most commercial meat-goat herds are located in the Southwest and South.

Prior to imports of Boer goats in the mid-1990s, meat-goat herds in the United States were largely made up of Spanish Meat goats and other landrace goats brought into the United States by settlers. US goat meat, other than in Texas, was largely the byproduct of the dairy goat industry through excess bucks and cull animals. By the 1990s, breeders looked to improve the yield and quality of their animals. This led to the importation of Boer goats from Australia and New Zealand directly into the United States or through Canada. Additional breeds such as the Savannah and Cashmere goats were imported and Myotonic goats were bred specifically for meat or mixed into meat herds.

Most meat-goat registries in the United States seem to be breed specific. As such, I will cover them under the individual breed descriptions below.

Boer

Boer goats were originally developed as a meat breed in South Africa. *Boer* means "farmer," the same name given to Dutch farmers in South Africa. This farmer's goat is known for heavy muscling and rapid weight gain.

Boers initially came to the United States by way of Australia, as the USDA restricted the import of goats directly from Africa. In the late 1980s, South African producers smuggled Angora embryos into New Zealand and Australia; almost as an afterthought, some Boer embryos were included. Two main farms, African Goat Flocks and Landcorp, then exported the very first Boers to the United States. Those first Boers arrived in 1993 from New Zealand at export costs of $8,000 to $10,000. What followed was a frenzy of exotic and speculator buying. One Boer buck sold for $80,000!

A number of registries formed with the arrival of Boer goats. The American Boer Goat Association (ABGA) formed in 1993 when the first Boer goats arrived in the US. ABGA continues to be the largest registry for Boer goats. The International Boer Goat Association folded at the end of 2012 and transferred its record database to ABGA. The US Boer Goat Association was formed in 1997 and also has a comprehensive registry.

When I attended the first American Boer Goat Association national convention in 1994, it surprised me that people could get so worked up about a goat. The bubble burst quickly. By mid-1995, the USDA had changed import rules to allow semen and embryos from South Africa via breeders in Canada, Australia, and New Zealand. That same year, some auctions were selling does for as little as $800. Some Boer breeders worry that the initial pampering of these rare, expensive animals affected their hardiness and growth. Today, Boers of acceptable quality can be purchased at the same cost as other goat breeds.

Confusion exists about the definition of a true full-blooded Boer goat. Regardless of the origin of import, all Boer goats tracing their pedigree directly back to South Africa are full-bloods. Percentage Boers are those with at least one goat of non-African breeding in the pedigree.

Kalahari Red

A new breed out of South Africa is the Kalahari Red goat. There is plenty of interest in the United States about this breed. Unfortunately, farmers in the United States cannot directly import animals from South Africa. The herds now being raised in Australia and New Zealand are probably going to be the initial countries of import similar to the initial introductions of Boer goats. Although this large red goat looks very similar to a Boer, genetic and blood testing from the Agricultural Research Council of South Africa has shown no genetic link. They are named after the Kalahari desert, which spans the borders of Botswana, South Africa, and Namibia. In this rough climate, they were largely developed through natural selection. These are particularly sun-, parasite-, and disease-resistant livestock. They also have strong herd instincts.

The main characteristic of Kalahari Red goats is their red coat color, which was developed because a lighter color does not camouflage well in their native land. Kalahari Reds have long, floppy ears and moderately sized, sloping horns. Body type is similar to that of the South African Boer goat. Kalahari Red goats can be used as a good crossbreed to increase hardiness and carcass size. Bucks should be larger than does, with loose skin in the neck region.

I predict that in the next few years we will start seeing these big red goats in US goat meat herds. Meanwhile, people new to goats are already looking for this breed in online forums such as Facebook. I'm sure with import costs, the early arriving goats will be quite expensive. A registry will most likely be created to track the breed. Meanwhile, breeders' efforts to breed an American Kalahari Red using red Boers crossed with other meat breeds have not been successful. If you find Kalahari Red goats for sale in the United States, be sure to check the provenance.

Kiko

The Kiko was bred by a New Zealand consortium of large goat farms. This group crossed feral does with Anglo-Nubian, Toggenburg, and Saanen bucks. Within four generations, a new breed of meat goat was born. It showed dramatic improvements in live weight. The new meat breed was christened the Kiko (for *kikokiko*, Polynesian for "flesh consumption"). By 1986, New Zealand breeders had closed the Kiko Herd Book. It is the only goat bred specifically for performance rather than appearance.

Efforts are being made to breed the Kiko true to its hardy origins. Known for rapid weight gain and foraging ability without supplemental feed, the Kiko is supposed to thrive under all conditions without human intervention. A study in 2004 at Tennessee State University seemed to agree. Their results seemed to show that Kiko goats had greater resistance to internal parasites and were less susceptible to foot rot than other breeds. There are three registries devoted to Kiko goats—the American Kiko Goat Association, the International Kiko Goat Association, and the National Kiko Registry.

Most Kiko goats are white with brown eyes and dark skin color, although the American Kiko Goat Association allows goats with other skin, hair, and eye colors to be registered.

Myotonic Goat

Myotonic, Tennessee Fainting, Tennessee Meat, Texas Wooden Leg, Stiff, Nervous, and even Scare—these names all refer to a goat breed with a condition called myotonia congenita. The muscle cells tighten when the animal is startled. Fainting goats sometimes fall down due to momentary stiffness in the legs. This action is unrelated to the nervous system or actual fainting. Some Myotonics become very stiff, while others barely stiffen at all.

In the 1880s, an itinerant farm laborer named John Tinsley came to central Tennessee. He had four unusual goats that would stiffen when alarmed. Fellow farmers noticed they were less apt to climb fences and escape from pastures than other goats. The goats' muscular conformation and high reproductive rate also caught breeders' attention. Soon, Stiff goats became known across the region. Tennessee Fainting goats spread to Texas in the 1950s. Further selection for larger size and improved meat developed a variety called Texas Wooden Leg goats. The ears of a Fainting goat look like a cross between Nubian and Swiss breeds. Essentially a landrace breed, Myotonic goats have always varied in size. There are two types of fainters: novelty animals selected for small size and stiffness and meat animals bred for double-muscling, rapid growth, and good reproduction. Mature weights range from 60 to 175 pounds.

Stiff-Legged goats are said to be calmer than other breeds, with strong parasite resistance and good mothering ability, in addition to having myotonia. Does produce multiple kids in a wide array of colors and patterns and freshen often, sometimes every six months. Occasionally, individuals will have blue eyes. They are excellent mothers with plenty of milk for twins or triplets.

Around 1980, Fainting goats were listed on the endangered/critical list published by the ALBC. Due to an increased interest in fainters, the breed is now listed as recovering. Myotonic goats are being crossed with other meat breeds, such as the Boer. While improving meat traits and breeding efficiency, this practice may be depleting the Myotonic gene pool.

Several registries serve Myotonic goat breeders. The International Fainting Goat Association (IFGA) registers Fainting goats in one of the four categories: Certified Premium, Premium, Heritage, or Regular.

CERTIFIED PREMIUM. This category is for all animals that readily faint and fall down in a myotonic state. This must be documented by photograph. Both the sire and dam must be registered with the IFGA or another registry.

All Fainting goats are heavily muscled because of the extra work these muscles do as they stiffen. Short and long-haired varieties exist with some producing cashmere. *Jen Brown, Hidden Springs Game Farm*

PREMIUM. This category is for all animals that readily faint and fall down in a myotonic state. This category does not require a down photo, but both the sire and dam must be registered with the IFGA.

HERITAGE. Goats that readily faint but do not meet the lineage requirements of the Premium category. This category requires a photo in a full down myotonic state.

REGULAR. Does that are wooden legged but don't fall over in a full myotonic state. Bucks cannot be registered in the Regular category.

The Myotonic Goat Registry (MGR) has a long, specific breed standard. However, MGR registers both purebred and percentage crosses of the traditional fainter. According to their website, "This would allow breeders to 'grade up' to purebred status, by clearly identifying goats that were known to be partly outside breeding. This way, even if a 75-percent Myotonic goat 'looked right,' it could not be misidentified as a purebred animal. Registering the crossbreds or upgrades in herds of serious breeders gives these goats a specific identity. This greatly reduces the chance that they or their offspring will be presented to the registry as purebreds." They will register goats that do not faint.

The American Fainting Goat Organization (AFGO) is a family-owned registry formed in 2011 to promote and keep the Fainting goat as close to the historical breed as possible. They evaluate each individual goat against standard and will not register a goat that does not meet standard. It is an open registry that will not record known crossbred animals. AFGO registers Fainting goats and miniature Fainting goats. Mini Fainting goats meet the Fainting goat standard, but are 22 to 23¾ inches and under at the age of three years old.

Pedigree International, established in March 2000, also registers Myotonic goats.

Savanna Goat

Around 1957, another breed of meat goat was being developed in South Africa by the breeder Cilliers and Sons along the Vaal River. Mev Cilliers received a white goat from her indigenous servants. She sought out five more goats with white in their color pattern. The white offspring of these animals became parents to a new breed. This part of the world is known for its rugged brush country, and the original herd of Savanna goats was left alone before and after kidding to allow natural selection to occur. This management practice created a hardy goat breed with good foraging ability and high parasite tolerance. The Savanna goat has short hair that develops a fluffy cashmere layer during the winter. Savanna goats are selected for all-black skin, which protects them from the harsh sun. Pedigree International registers only pure white goats as Savannas. Those Savannas showing color, but being 100 percent Savanna, are registered as "American Royal."

Pedigree International is the first registry for the Savanna goat in the United States and is the largest database in North America for the Savanna goat. Pedigree International is also the original registry for Cashmere goats. It is the only registry of Spanish, TMG, and TexMaster goats. Pedigree International maintains herd books for Kiko, Gene-Master, and Myotonic goats as well as offering private farm pedigree services in addition to USDA scrapie pet and pack goat registration.

Spanish Meat Goat

During the sixteenth century, landrace goats from the Mediterranean traveled with Spanish missionaries to the Caribbean. Those travels led to the Americas, where the Spanish goats flourished. For three hundred years, these were the only goats in what would become Mexico and the United States. Meanwhile, the foundation stock in Spain died out, making the Spanish goats overseas of special importance both genetically and historically. In those early days, the Spanish goat was valued for both milk and meat.

Many people consider the Spanish goat not a true breed, but rather brush or scrub goats. These "woods" goats actually include both crossbred and purebred animals. Breed conservancy is difficult because of the wide variability in physical characteristics in regional herds, which lack registries or lineage records. The term *Spanish goat* is used to describe almost any Southwestern meat goat of unknown parentage.

An increased interest in meat goats has led to renewed interest in Spanish goats. These goats thrive in the toughest environments. Texas is home to large herds of Spanish goats valued for their longevity and fertility. Goat size varies, with adult weights ranging from 50 to 200 pounds. Their ears may be as long as those of the Nubian but are held horizontally next to the head in a distinct manner.

Pure strains are being diluted as breeders cross Spanish goats with Boers or Kikos for hybrid vigor and increased meat production. Cashmere breeders are also trying to crossbreed with Spanish goats for better fiber production. Even with an estimated population as large as eight thousand animals, conservationists have the Spanish goat on watch status. Attempts are underway to preserve a pure strain.

The Spanish Goat Association (SGA) was started in August 2007 through the work of Justin Pitts and the American Livestock Breeds Conservancy. SGA does not keep a registry. It lists bloodlines and herds after "painstakingly verifying them as Spanish by history, visual evaluations, and the trusted 'word' and reputation of the breeder." In 2017, as the number of identified Spanish goats quadrupled, the SGA decided there needed to be support for better breed tracking. The nonprofit Spanish Goat Registration Society was formed to safeguard the genetic diversity of Spanish goats.

On the cutting edge of science, DNA collection is hoped to provide data to track and monitor Spanish goats. Current fears are that modern meat-goat management methods will dilute or even erase Spanish bloodlines through culling and crossbreeding. With the help of the University of California—Davis, data collected by this society is helping to create markers to determine Spanish goat DNA. Membership in the Spanish Goat Registration Society classifies herds as "Verified" or "Registered." Registered Spanish Goat Herds are those that allow a number of their goats to be DNA tested for the UC Davis database. It does not require that all goats in a herd be registered. This was deemed impractical for large herds or commercial meat producers to include all goats due to the expense, time, and energy involved. A Verified Herd is one that does not involve DNA testing.

MINIATURE GOATS

Miniature goats can be used for many of the same purposes as their full-size counterparts, including milk, fiber, and meat. However, the vast majority of miniatures are kept as pets and exotic animals. As discussed under Dairy Goats above, the Nigerian Dwarf is a miniature breed. Other minis have generally been created through crossbreeding either a Pygmy or Nigerian Dwarf bucks to a full-size doe from a variety of types and breeds. It is not recommended to breed a miniature doe with a full-sized buck because of potentially oversized kids risking the life of the smaller dam.

Pygmy

In 1959, the first documented African Pygmy Dwarf goats arrived in the United States from Sweden, where they had been novelty animals in zoos and exhibits.

They were called Cameroon Dwarf goats after the former French Cameroon area in West Africa, where they originated. The West African Dwarf goat is widely raised for meat in its native land; the population is thought to be over sixteen million head! Meanwhile, Pygmies have spread across the United States as pets, exotics, meat, and laboratory animals. Their popularity as pets derives not only from their size, but also from their friendly, playful manner and gymnastic antics.

Unlike the Nigerian Dwarf, which has a proportional body type, the Pygmy is an achondroplastic dwarf. Often displaying bowed legs, this charming miniature has a disproportionately large body compared to the length of its legs. This is also described as "cobby." The trait was probably an adaptation to the hot West African climate. The recessive genes for dwarfism led to animals that could tolerate the high temperatures better than their larger and more proportional herd mates.

The Pygmy's hair varies in length depending on the climate. The does do not develop much of a beard, but the bucks have flowing beards and handsome shoulder capes. The Pygmy comes in multiple body colors. As of 2013, the National Pygmy Goat Association lists seven color choices: caramel with black markings, caramel with brown markings, gray

Available in many colors, the Pygmy is popular for its friendly, playful manner and gymnastic antics. *David Weber, Cutter Farms*

agouti, brown agouti, black agouti, black, and solid black. All agoutis have solid stockings darker than the main body color, with black stockings on gray agouti and brown stockings on brown agouti. Muzzle, crown, eyes, and ears are distinctly accented in white and may be intermingled with hairs in the same color as the body.

The breed standard for registered Pygmies requires a darker body, crown, dorsal stripe, martingale, lower legs, and hooves against a lighter muzzle, forehead, eyes, and ears, except in the case of all-black animals. Caramel-colored goats also need vertical stripes on the front side of darker socks. Male Pygmy goats should be 16 to 23⅝ inches measured at the withers, and the females are slightly smaller at 16 to 22⅜ inches.

Mini Dairy Goats

One of the first official miniature versions of a larger dairy breed, the Kinder was created by crossing a registered Pygmy buck to Nubian does. The first three kids, all does, were born in 1986. The resulting airplane-eared offspring were then back-crossed with one another.

Following the Kinder, breeders created minis of most dairy breeds by both accident and design. Much of the experimentation may have been due to more dairy goat farmers branching into the pet breed niche and raising both types of goats. Regardless of what gave breeders the idea, the minis are here and popular with people who want a unique pet with dairy characteristics. Miniature goats produce plenty of milk for the smaller modern family and require less space and feed.

True to the popularity of miniature goats, there are multiple registries for these compact animals. The Miniature Goat Registry (TMGR) and the Miniature Dairy Goat Association (MDGA) both provide registration services for Experimental, American, and Purebred Miniature dairy goats. They have specific breed standards for mini goats that arose from crossing Nigerian Dwarf goats with the full-sized goats of recognized dairy breeds. The breeds are Mini Nubian, Mini LaMancha, Mini Alpine, Mini Oberhasli, Mini Saanen, Mini Sable, and Mini Toggenburg. All of these breeds may have blue eyes due to the influence of Nigerian genetics.

The International Goat, Sheep, and Camelid Registry, as mentioned in an earlier section, registers mini goats of all types.

RARE BREEDS

Goat biodiversity holds special interest for conservationists and historians. Across all species, the world loses an average of two domestic animal breeds each week, according to the Food and Agriculture Organization (FAO) of the United Nations.

The concern about losing genetic variation among livestock is real. Rare breeds carry traits for disease resistance, mothering ability, and weather tolerance that can be lost in commercial herds bred for conformity and production. Problems also occur when commercial demands spread the animals to environments less suitable than their native climates. Uniformity of type may be convenient for large-scale farming, but it creates vulnerabilities. In a worst-case scenario, a dominant commercial breed could catch a deadly disease. Such a calamity could wipe out the group.

The ALBC works to conserve historic breeds while supporting genetic diversity in livestock. This group has six goat breeds on its Conservation Priority List. The two rarest are the Arapawa and the San Clemente, landrace goats that developed in the isolation of island homes after being deposited there during colonial expansion from Spain and England. Other listed breeds are the Spanish, the Myotonic, the Oberhasli, and the Golden Guernsey.

A smaller breed, Arapawa goats are lean and have light-boned frames. They vary in color, although most registered animals are white, gray, or black mixes. *Al Caldwell, Long Path Farm*

Arapawa

Olde English Milche goats arrived in the South Pacific with Captain James Cook and other colonial explorers in the late eighteenth century. One such location was Arapaoa Island (formerly Arapawa Island) in the Marlborough Sounds of New Zealand, where a group of feral goats flourished for two hundred years. Now extinct in their native England, these goats recently faced extermination at the hands of New Zealand authorities, who considered the herd nonnative vermin. The controversy caught the attention of Betty and Walter Rowe, who had moved to New Zealand from the United States in the 1970s. Their efforts led to the formation of the Arapawa Wildlife Sanctuary in 1986, which continues to preserve the breed today.

The New Zealand Arapawa goat or Arapawa "Island" goat is one of the rarest goats in the world. The ALBC has listed the Arapawa status as "critical." In 1993, a founding US herd was established at Plimoth Plantation, a living history museum in Plymouth, Massachusetts. These six animals form the entire genetic stock of Arapawa goats in this country. The Sedgwick County Zoo in Wichita, Kansas, has a small herd in its petting zoo. Only about fifteen total farms across the United States maintain breeding herds.

Recent studies conducted by Dr. Phillip Sponenburg of the ALBC showed the Arapawa to be a unique genetic breed. DNA studies found that they are definitely not Spanish goats. Historical documentation still leans toward them being one of the last remnants of the Olde English Milche goat, but further study is needed. Current numbers of Arapwawa goats in the world are estimated to be a little over 400 recorded animals.

Golden Guernseys come in any shade of gold and may have some white marking. They are smaller in size and less wedge-shaped than other dairy goats. *Phyllis Clayton, Golden Guernsey Goat Society*

Golden Guernsey

A minority breed even in England, the Golden Guernsey has long been coveted by American goat fanciers. The import of these animals is presently illegal. Guernsey-type goats are being bred in the United States using Golden Guernsey semen crossed with Swiss dairy goats. The only true Golden Guernsey, however, is one registered in the British Golden Guernsey Herd Book.

Golden-haired dairy goats are mentioned in English records as early as 1826. The goats initially arrived with trading ships. The Golden Guernsey lines most likely came from the Maltese Islands, with origins in Greece and Syria. Starting in the 1920s, they were registered in the English Guernsey Herd Book. By 1965, a distinct group of golden goats was breeding true, and a separate Golden Guernsey Herd Book was started. This closed registry allows no grading up from crosses with other breeds.

While there are physical similarities to Swiss breeds of goat, the breed's golden skin and genetic differences make this a distinct breed. The ears on this medium-sized goat are erect and often set slightly lower than Swiss breeds and carried horizontally, or forward, in what is termed the "bonnet" position. The Guernsey coat or hair color can be any shade of gold, but the skin color must display some gold tone, ranging from peachy flesh to orange-gold with gold-toned skin preferred.

Southwind Farms in New York had the first herd of purebred Golden Guernseys in the United States. Their first embryo-offspring were born in 1998 after the import of embryos through Canada a few years earlier. Unfortunately, import of embryos is now prohibited. Another group of US breeders used Guernsey semen to upgrade their dairy does of other breeds to Guernsey genetics. In 2011 and 2015, the Guernsey Goat Breeders of America petitioned ADGA for the registration of Guernsey goats as a recognized dairy breed with specific breed standards. Survey numbers listed on the ADGA website indicate that about 73 percent of ADGA members responding to a survey support this breed's inclusion into the registry.

San Clemente

Off the coast of southern California lies San Clemente Island, home to a landrace breed of goats that inhabited the island following visits by colonial ships. Similar to the Arapawa Island goat, San Clemente goats were not indigenous to their island and damaged the native flora and fauna. However, their isolation has also given them unique genetic markers which show these goats to be a distinct species. Their conservation status is critical.

The US Navy has been responsible for San Clemente Island since 1934. Locked onto the island, the San Clemente goat population soared. By a 1972 survey, at least fifteen thousand goats were living on 57 square miles. The goats destroyed vegetation and endangered the ecosystem. In the 1970s, San Clemente Island was made a preserve for native species. Since the goats were not native, they had to go.

The navy first used hunting and trapping to eliminate the feral goats. The animal protection group Fund for Animals spearheaded a program of systematic removal, which saved certain animals from destruction. Bucks were neutered and females adopted out with "no breed" clauses. By 1991, no goats remained on San Clemente Island.

Slightly larger than miniature breeds, San Clementes have a deerlike appearance and a variety of colors predominated by red or tan with black markings. The head is long and lean with a dished appearance. Narrow ears are carried horizontal to the head and have a distinct crimp in the middle. Both sexes have horns that resemble those of Spanish goats. However, DNA studies have determined that the San Clemente breed is distinct from Spanish goats.

The San Clemente Island Goat Association is working to increase the geographical distribution and preserve the genetic diversity of San Clemente goats. Only about 750 of these animals remain today, with fewer than forty breeders worldwide. Animals are registered with the International Goat, Sheep, and Camelid Registry. The San Clemente Island Goat Foundation is also committed to preserving this historic animal.

A PHYSICAL AND BEHAVIORAL OVERVIEW

Once you learn the many parts of a goat, you'll be able to master the quiz bowls put on by 4-H and the National FFA Organization. There are also more practical reasons for goat breeders to be familiar with these terms. I have learned quite a bit about good conformation by watching judges or appraisers evaluate animals. Until I learned the language, their comments meant very little. Now I can appreciate a goat that is "higher in the chine" or "wider in the escutcheon." By studying the charts that follow, you can also learn to evaluate your goat.

MOUTH

The goat's mouth has both a hard palate and teeth. There are no teeth on the top front. When goats browse, they use the bottom teeth and the hard upper gum pad to break off vegetation.

Just because you don't see opposing teeth doesn't mean there aren't any. In the back of the goat's mouth are grinding teeth. Any child who has managed to stick his hand far enough into a goat's mouth will tell you that these teeth can hurt! The goat passes food by the tongue into the back of the mouth, where the opposing teeth grind it into pieces.

It is important that the teeth meet the palate correctly in the mouth. Two teeth-related congenital conditions can be found in goats. Parrot mouth occurs when the top jaw is longer than the bottom jaw. The other condition,

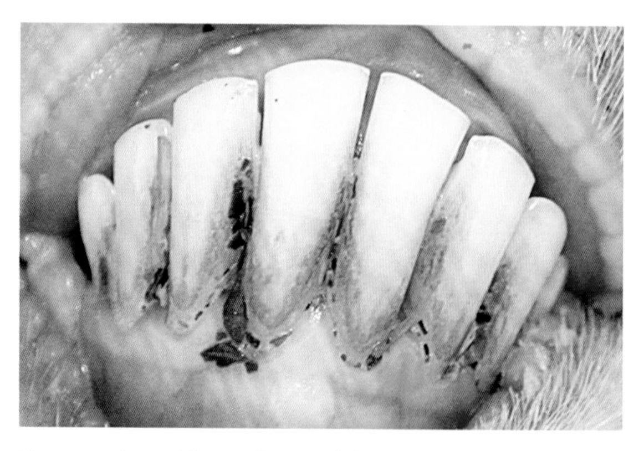

The upper jaw of the goat's mouth has no front teeth. Each year, baby teeth on the bottom jaw are replaced by larger adult teeth. By the age of four years, all eight incisors are usually mature and the goat is now known as an "eight tooth." *Jen Brown*

undershot or monkey mouth, occurs when the lower jaw is longer than the upper jaw. Either condition is grounds for culling, as each makes it more difficult for the goat to collect and chew browse, which can reduce feeding efficiency and milk production.

The front teeth can be a good indicator of a goat's age. At birth, there are six incisors on the bottom jaw.

Early in the goat's life, these teeth are just starting to break through the skin. By four weeks, there are normally eight incisors, known as milk teeth, in the front of the lower jaw, as well as twelve molars in the back of both the top and bottom jaws. The teeth wear with age, so the older the goat, the more worn down the teeth become. The teeth also spread in the mouth, become loose, or fall out. A goat that has lost all its teeth is known as gummy.

DIGESTIVE SYSTEM

As a ruminant, the goat has a stomach system made up of four compartments. The first two chambers are the rumen and the reticulum (also known as the honeycomb), which work with saliva and stomach juices to break down fiber. The final two chambers are the omasum and abomasum, or true stomach.

In kids, milk bypasses the other chambers to be digested directly by the abomasum. Stomach acids in this final chamber digest the milk without the need for fermentation to break down roughage. At this stage of development, the abomasum is the largest chamber. As the kid starts eating roughage, the rumen and other chambers become more active. By adulthood, the rumen has grown to be the largest chamber.

PARTS OF THE GOAT

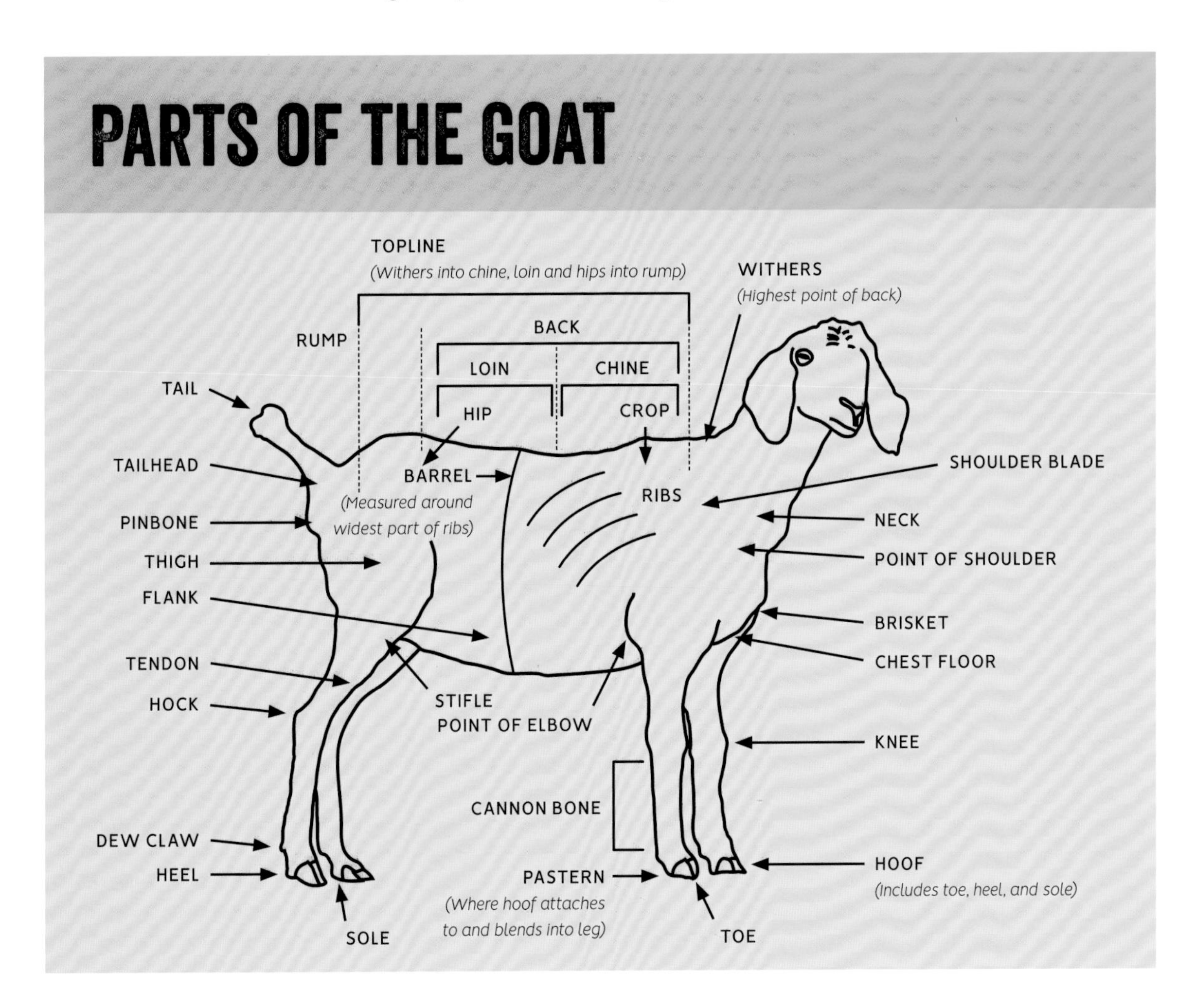

PARTS OF THE MAMMARY SYSTEM

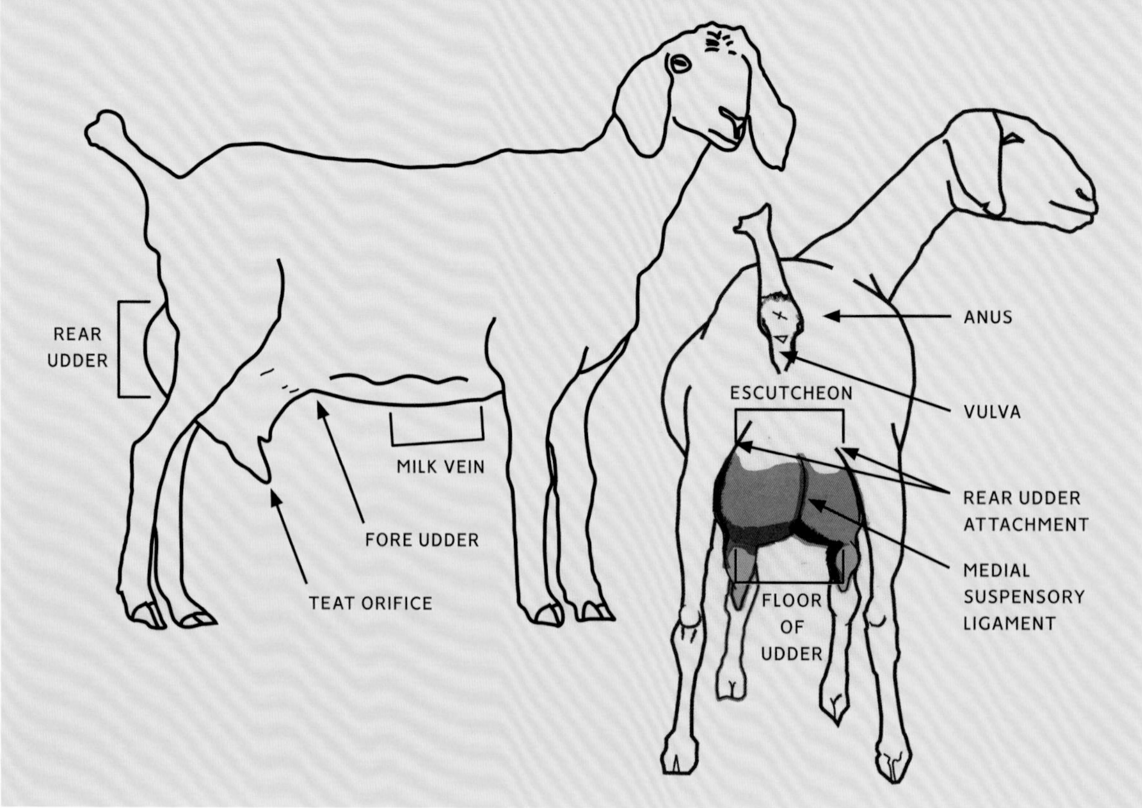

At fairs, I am often asked if one of my older milkers is pregnant. I explain that the large, round belly is caused by her rumen. This big fermentation chamber allows the ruminant to break down cellulose into digestible materials with the help of enzymes and bacteria. The goat chews roughage and swallows it into the rumen. Once it is mixed with digestive juices, the mixture is regurgitated in the form of cud. Cud is chewed and swallowed multiple times and passed between the rumen and reticulum until it is semi-liquid. The finely chewed product is then swallowed again and passes into the next stomach chamber—the omasum. The omasum is where much of the water and mineral nutrients are absorbed into the bloodstream. Finally, the semi-processed food passes into the abomasum. There, additional digestion takes place before the contents pass into the intestinal system.

BODY CONDITION SCORING

Body condition scoring is a system used to evaluate the amount of body fat on an animal. The score indicates whether the goat is skinny, just right, or fat. A 1 to 5 classification system is the most common, with 1 being very thin and 5 being very fat. Most goats thrive in a body condition score 3.

BASIC GOAT BEHAVIOR

The images in Carmen Bernos de Gasztold's poem "Prayer of the Goat" (page 42) are right on the money! She emphasizes that goats are not sheep.

PARTS OF THE BUCK

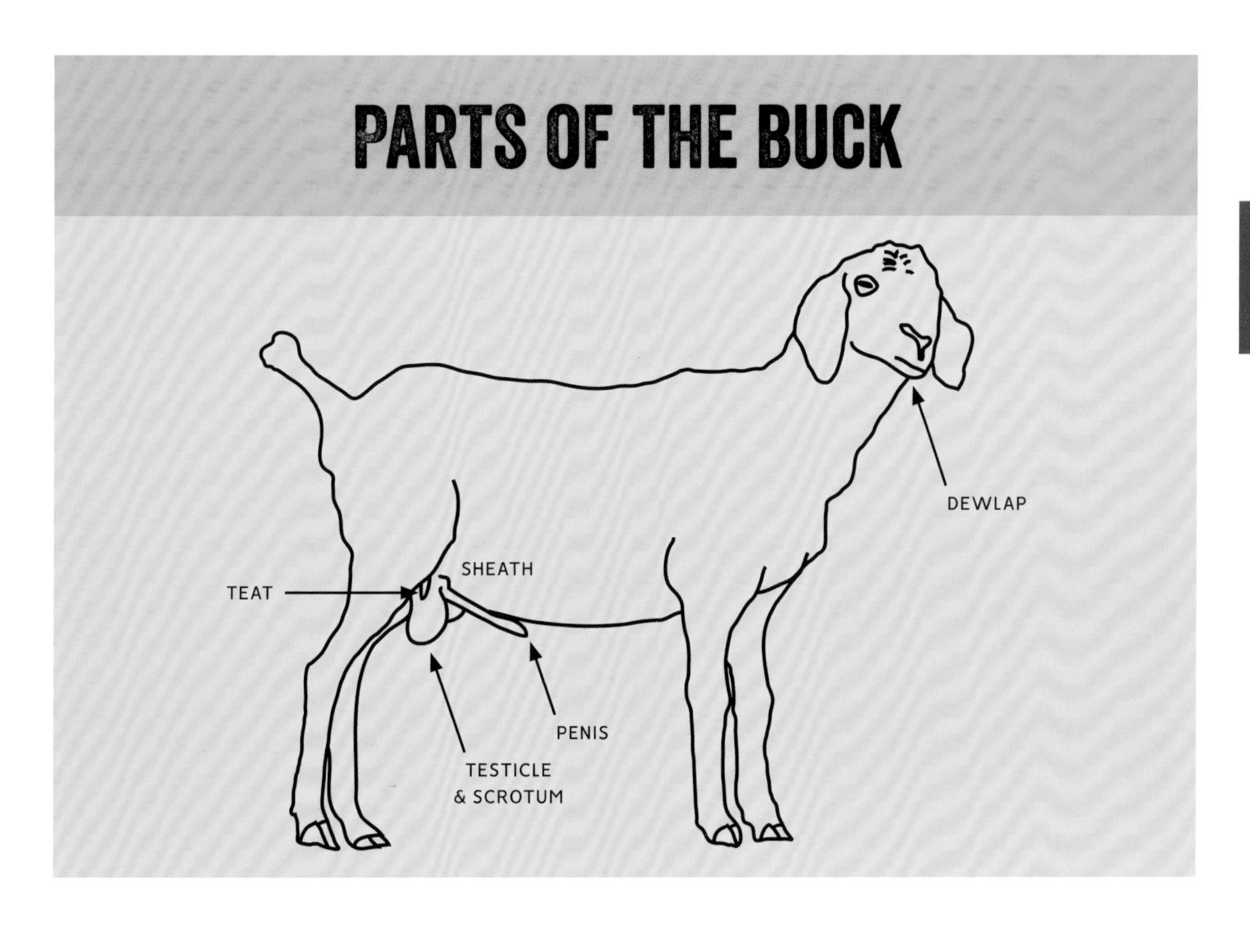

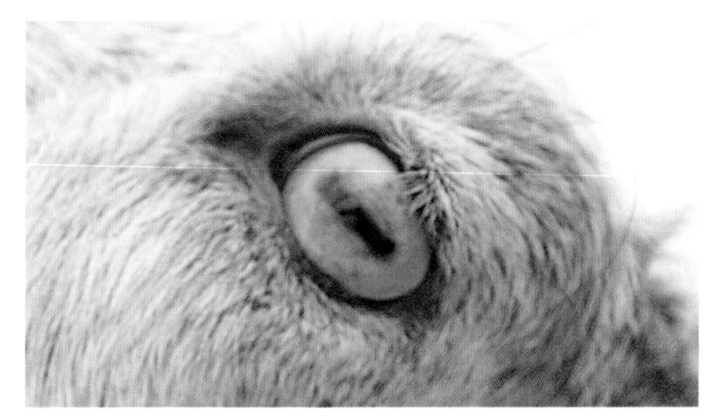

The rectangular shape of the pupil gives goat eyes a unique appearance. Most goats have brown eyes (above), although some Nigerians have blue eyes (left). *Jen Brown*

The layperson tends to lump sheep and goats together. Both species are ruminants. Both provide meat, milk, hair, and hide. But there are key differences. Sheep and goats don't eat the same way. Sheep graze, feeding almost exclusively on grass and plants growing on the ground. Goats browse, searching around for the most tender vegetation, wherever it might be. They nibble at the leaves from the bushes next to their pen, stand up on their hind legs to reach tree branches, and look to the choice new-grown blossoms of clover in their pasture. The false statement that goats eat anything—or worse, that goats eat garbage—stems from their natural tendency to browse. Their inquisitive lips reach out to taste anything that catches the eye. The paper wrapping on a tin can is made from wood products. That paper, combined

HEAD OF THE GOAT

POLL
EAR
FOREHEAD
EYE
BRIDGE OF NOSE
NOSTRIL
MUZZLE
JAW
THROAT
DEWLAP
WATTLE

Goats, like cows, raise and lower themselves by bending their front legs first rather than their back legs.
Jen Brown

with the glue on the can, is very tasty. But eating it creates the impression that the goat wants to eat the tin can as well!

Goats are not as flock-oriented as sheep. Watch goats in the pasture. The herd spreads out, wandering. The animals go a set distance before they become too uncomfortable to separate farther from their herd mates. The lead goat is kept in sight, and that goat's actions influence the others. Sometimes, a small group of goats—or a loner—leaves the main herd to go adventuring. I have chuckled at the cry of distress and panicked running when a goat discovers that the herd has moved on while she had her nose buried in a bit of forage. I once picked up a "problem" goat. The owners were beside themselves because the solitary goat had destroyed their lilac bushes during her constant escapes. Penned with other goats, she settled right down. In the absence of other goats, a lone goat will look for company. The animal will escape its pen and cry plaintively. Then the stereotype of the goat in the clothesline and climbing on the car becomes reality.

Another crucial difference between goats and sheep is their tolerance of the elements. Sheep have wool that repels water, allowing them to stay out in all sorts of weather. Goats hate to get wet. Even woolly Angora goats need protection from the elements. The lack of a water-repellent coat makes goats prone to chilling and pneumonia. It is easy to tell when rain is imminent, because the goats disappear from the pasture.

Cud is chewed using the back teeth. *Jen Brown*

The ADGA has a formal process for evaluating dairy goats known as Linear Appraisal. Senior dairy goat judges trained as evaluators come to your farm and assess the animals using a standardized format that gives breeders a measure for breeding goats approaching a certain type. Other appraisal systems are in place with various breed associations. *Carol Amundson, Terrapin Acres*

PRAYER OF THE GOAT
BY CARMEN BERNOS DE GASZTOLD

Lord, let me live as I will!
A little giddiness of heart,
the strange taste of unknown flowers.
For whom else are Your mountains?
Your snow wind? These springs?
The sheep do not understand.
They graze and graze,
all of them, and always in the same direction,
and then eternally
chew the cud of their insipid routine.
But I—I love to bound to the heart of all
Your marvels,
leap Your chasms,
and, my mouth stuffed with intoxicating grasses,
quiver with an adventurer's delight
on the summit of the world!

This underconditioned milker has very little fat on her body, earning a body condition score of 1 to 2. *Jen Brown*

This overconditioned wether has quite a bit of fat under his winter coat, earning a body condition score of 4 to 5.

A playful goat bounds and leaps and quivers.
Barb O'Meehan

SCORE	CONDITION	BACKBONE AND RIBS	LOIN
1	Very Lean	Easy to see and feel Can feel under ribs	No fat
2	Lean	Easy to feel Smooth Need to use a little pressure to feel ribs	Smooth fat
3	Good	Smooth and rounded Even feel to the ribs	Smooth fat
4	Fat	Can feel backbone with firm pressure No points on spine and no ribs felt Indent between ribs felt with pressure	Thick fat
5	Obese	Smooth No individual vertebrae felt No separation of ribs felt	Thick fat Lumpy Jiggles

Goats browse on a variety of plants. More like deer than sheep in their feeding habits, goats like to pick and choose their food. *Jen Brown*

Although less flock-oriented than sheep, a lone goat often panics once it discovers it has become completely parted from the herd. *Barb O'Meehan*

Goats rear high in the air to confront attackers. *Jen Brown*

Goats	Sheep
Males are called bucks	Males are called rams
Females are called does	Females are called ewes
Bearded	Beardless
Tail held upright	Tail hangs toward the ground
60 chromosomes	54 chromosomes
Browsers	Grazers
Maintain some independence within a herd	Flock closely together
Susceptible to rain	Fleece protects from rain
Butt downward after rearing on hind legs	Butt charging straight ahead without rearing

2 PREPARING YOUR GOAT FARM

DON'T APPROACH A GOAT FROM THE FRONT, A HORSE FROM THE BACK, OR A FOOL FROM ANY SIDE.

—YIDDISH PROVERB

Goats don't require a lot of space unless you raise them on pasture or grow their feed yourself. A goat dairy on 5 acres might have more than one hundred milkers. Redwood Hills Dairy in California keeps more than three hundred goats on 10 acres. Its facilities include open housing with little pasture beyond a lounging yard. Although this lifestyle may sound unappealing, goats actually prefer it. Even with access to outdoor space, dairy goats frequently lounge around the barn, enjoying the close proximity to water and feed. In fact, Redwood Hills is the first goat dairy in the United States to earn Certified Humane® farm status. Clearly, these goats are neither overcrowded nor abused.

Above: Ear tags are required by some feedlots, even with the presence of tattoos or other ID. *Jen Brown*

Opposite: Goats often prefer to stay near the barn, close to their source of food and water. *Jen Brown, Poplar Hill*

Pet goats are usually kept in smaller numbers and housed accordingly. A large dog kennel can serve as caprine housing if the animal has plenty of exercise time with the owner. Meat- or fiber-goat operations frequently run on range or pasture. Range requires the most land. Stocking rates of six to twelve goats per acre are recommended. The type and amount of forage available to the animals, as well as the weather, create many variables in the number of goats that a tract of land can support. A certain amount of feed and housing supplementation is usually required during harsh weather and during the breeding season.

REGULATIONS

You are more likely to encounter restrictions on your goat herd due to municipal zoning regulations rather than personal space limitations. Larger towns and cities have specific rules against livestock. Even in the country, subdivision covenants often regulate livestock. Goats are often restricted to areas zoned as agricultural. Other regulations include acreage minimums, such as 5 acres minimum for keeping farm animals, and density limits, sometimes called "animal units," such as no more than ten mature goats (one animal unit) per acre.

With mad cow disease, avian influenza, and other bio-security concerns, premises identification is a hot issue for goat keepers. To prevent the spread of diseases, national and state regulations often require health tests and animal identification. Some states do not allow livestock to be exhibited at shows unless the farm has a unique premises ID. Once your farm has this code, you must keep a record of all animals entering and leaving the premises. ID systems typically require all animals over a certain age to be permanently identified. If you find yourself bringing your goats home from another state, be prepared. Interstate movement of animals is regulated.

Goats can lose ear tags to a variety of factors, including thinness of the ear, browsing in areas that catch the tag, the curiosity of other goats, and infections. *Jen Brown*

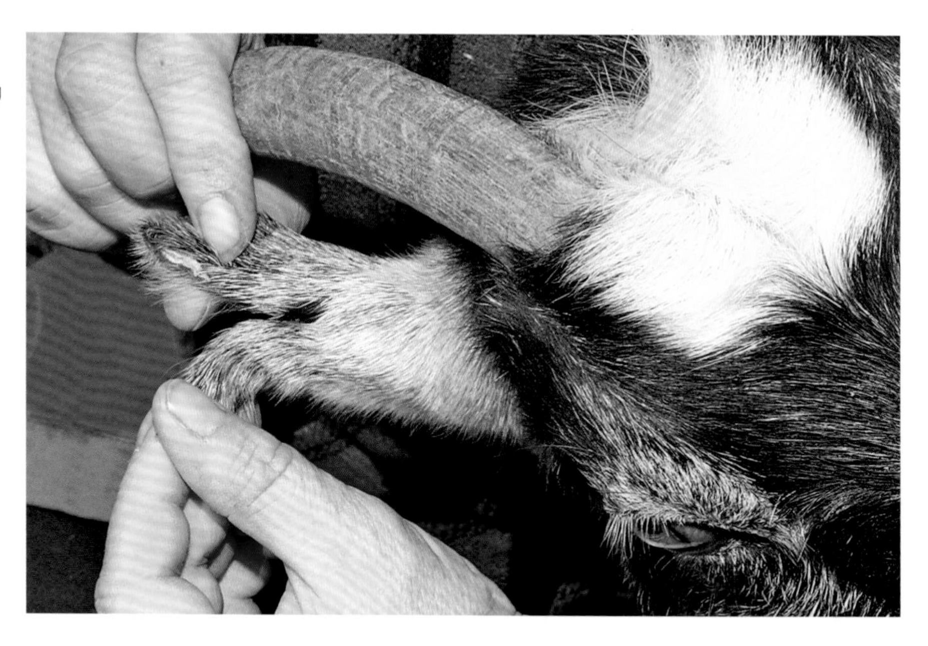

The National Scrapie Eradication Program, coordinated by the USDA's Animal and Plant Health Inspection Service (APHIS), dictates the identification methods approved for goats. Some states have requirements that are stricter than the federal requirements. It is prudent to check with a veterinarian or contact your state veterinarian's office for advice before moving your animal across state lines. Generally, there are approved ID methods, including eartags, tattoos, and/or microchips.

Most individual states also require a certificate of veterinary inspection, also known as a "health paper," for movement of livestock across state lines. Testing for tuberculosis, brucellosis, bluetongue, and other diseases may be needed. Animals from areas in which certain diseases are endemic, such as tuberculosis, may be prohibited from entering a state.

APPLYING EAR TAGS

1. Select the tag and record which goat you are tagging.
2. Wipe any dirt off the ear with a disinfectant, such as Nolvasan Solution or rubbing alcohol.
3. Load the tag pliers with the tag.
4. Dip the tag and the end of the pliers in a container of disinfectant.
5. Select a location on the ear away from obvious blood vessels and away from the edges to minimize tear losses.
6. Attach the tag to the goat's ear.

ANIMAL DISEASE TRACEABILITY AND THE NATIONAL ANIMAL IDENTIFICATION SYSTEM

The National Animal Identification System (NAIS) was a proposed system designed to track specific types of animals and diseases. Highly controversial, NAIS was dropped in September 2010 when comments to USDA made it apparent that it would be difficult, if not impossible, to implement. Scrapie eradication rules as well as state regulations are still in effect. Every goat breeder should be aware of these

TATTOOING

SUPPLIES

Tattoo pliers and characters
Restraining device
Alcohol pads
Toothbrush
Tattoo ink
Baking soda

1. Load the tattoo letters/numbers into the pliers. Check that they are correct by punching a piece of paper.
2. Have a second pair of pliers loaded for the other ear, or have your new letters/numbers set aside to change the set for tattooing the other ear or side of the tail.
3. Place smaller animals on your lap, between your knees, or in a disbudding box. Having an assistant can be very helpful. Place older animals in a stanchion.
4. Clean the ear or tail completely, using alcohol. Clip the ears if desired.
5. Apply ink to an area slightly larger than the tattoo on a flat section of the ear or tail between the ribs of cartilage.
6. Position the tongs. The bottom of the characters should be aligned facing the bottom of the ear or the outside edge of the tail.
7. Squeeze firmly and release. Be sure to lift the tongs out straight so that the tattoo doesn't get distorted.
8. Reapply the ink and rub in firmly with an old toothbrush.
9. While not necessary, rubbing baking soda onto the tattoo can improve the set of the ink.
10. Sanitize the tongs and characters between goats by dipping in alcohol or disinfectant.

rules and able to apply permanent identification to their animals as well as keep records of where they acquire or sell individual goats.

IDENTIFICATION

Permanent goat identification is used for registry and documentation purposes. The dairy-goat industry uses tattooing. The use of microchip identification is slowly increasing in the United States, and a number of registries have room on their animal ID forms to include a microchip number. The meat- and fiber-goat industries more commonly use ear tags. Other methods include ear notches and freeze branding. In most instances, transported goats, sales (especially auctions), and exhibits require permanent ID.

TATTOOING

A tattoo is a series of numbers and letters inked into the ear or tail web of the goat. It is one of the most permanent forms of ID. Your registry gives your herd identification letters; if you do not register, you may use your farm or personal initials. The herd ID goes on the right ear or tail web, and the individual animal ID goes on the left ear or web. For the left ear or web, a birth year letter identifies the birth year of each kid born on your farm. The letters *G, I, O, Q,* and *U* are usually not used as birth letters because they are hard to differentiate. Next to the year letter, put a number for the animal that identifies it within the rest of the year's crop. Most breeders start with 1 and work their way through the birth order. So if you've selected the letter *L* to stand for 2019, *L12* in the left ear of a dairy goat means it was the twelfth goat born in 2019. Be aware that some other goat registries have different letters for the years than dairy goat organizations. As a result, in the above example, *J12* in a Boer goat would indicate the twelfth animal born in 2019.

MICROCHIPPING AND ELECTRONIC IDENTIFICATION

Identification by radio frequency, also called RFID (radio frequency identification), EID (electronic identification), or simply "microchipping," has been around for more than fifty years. It began in World War II when radio signals were used to ID friendly aircraft. Radio waves are read and capture information stored on a tag or in a chip attached to an object or animal. By 1983, biochips were used for tracking in fisheries. Since then, government agencies, zoos, pet owners, and other groups have been tagging not only fish, but dogs, cats, livestock, lab animals, wildlife, clothes, cars, hazardous waste, and even employees. Millions of animals have been chipped. Most, if not all, veterinarians use microchip readers as a regular tool in their practice.

The USDA published final rules in 2013 for tracking livestock moving interstate. Certain methods of animal identification are acceptable, including a fifteen-digit ISO-compliant microchip or EID. The International Organization for Standardization (ISO) is a worldwide federation of national standards bodies from more than 160 countries, based in Geneva, Switzerland. Established in 1947, ISO has a standard identification system for international commerce. Official ID must begin with the designated US prefix 8-4-0 and come from officially approved USDA manufacturers. RFID is mandatory for goats in Canada and the European Union.

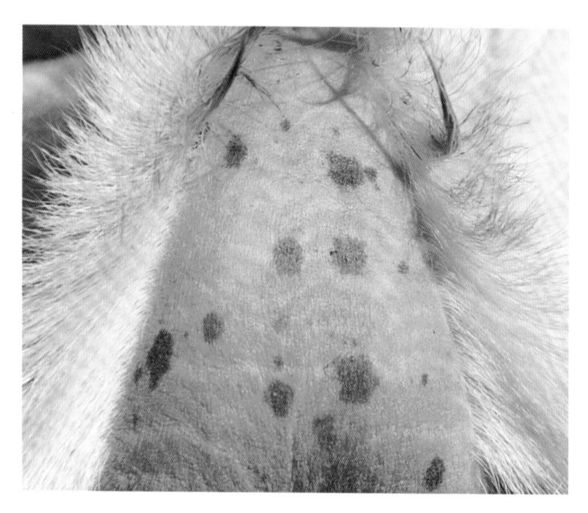

LaManchas are tattooed in the tail web because their ears are too small. *Jen Brown*

As of January 1, 2018, the ADGA allows a microchip as supplemental ID when recorded with the goat's registration papers. There is talk about making EID an acceptable independent form of ADGA identification. Other goat registries are increasingly adding microchipping to their ID forms and policies, more and more often as the only form of ID. Check with your particular breed registry if you wish to pursue this method of identification.

Microchip ID is more reliably permanent and has less chance of being lost than an ear tag. It is easier and quicker to apply, with less trauma, than tattoos in both ears or both sides of the tail. At future goat shows, I can see judges simply waving a reader and recording the information. No more arduous restraining of goats to read unreadable ear or tail tattoos. As a bonus, some laboratory animal or livestock microchips now read the animal's temperature. While chip-read temperatures aren't as accurate as the traditional rectal thermometer, it would be an additional tool for monitoring goat health.

Electronic tags are sometimes combined with visual tags such as ADGA's tattoo or an ear tag. The visual tag can serve as a backup ID. A chip reader is required to identify a microchip. These readers operate from 2 inches to more than 3 feet away from the animal, thus reducing handling of animals or cleaning tags, tails, or ears. Handheld stick or wand style readers are usually battery powered and Bluetooth enabled. For record-keeping or management, many software packages and mobile apps are able to accept RFID.

The microchip itself is about the size of a grain of rice and is encased in biocompatible glass or polymer covered by a sheath to prevent migration. It is inserted beneath the skin using a syringe in a manner similar to giving a subcutaneous (SQ) injection, although the needle is larger. Insert the microchip either at the base of the ear or in the tail web (the loose, hairless area under the tail on either side of the anus). You should choose a consistent spot on all goats in your herd to make finding the microchip easier. Be aware that chips have been known to migrate away from the original injection site. If your microchipped goat goes to slaughter, be sure to remove the chip or notify the butcher or inspector.

Microchips are often implanted by a veterinarian. Like other identification systems, microchips, injectors, and scanners are available to breeders as well. Be aware that, while easy to perform, chip injection needs to be precise so the chip is properly inserted between the skin and tissue in order to avoid complications or inability to be read. Follow your chip manufacturer's instructions precisely.

MANAGEMENT SYSTEMS

There is a spectrum of methods for raising goats, each with its own advantages and disadvantages. Assess your own personality and resources in order to decide how to proceed with your herd. Conventional caprine management techniques offer the widest number of options. Advantages include easy access to supplies and feeds, plenty of documentation and resources on techniques, and often lower supply costs. Many of these techniques were designed for large operations. Large herds may use a regular schedule of antibiotics, wormers, or chemicals to prevent diseases and pests. This routine may prevent health problems before they start brewing.

Disadvantages of conventional management exist in the same areas that make it advantageous. Routine use of antibiotics and wormers leads to bacterial and parasite resistance. Confinement housing can lead to higher disease and parasite loads. Drugs and chemicals that are used unnecessarily create greater expense.

The opposite of conventional husbandry is organic management. True organic livestock is difficult and costly to produce. Pasture, hay, and grain fed to organic goats must be free of antibiotics, chemicals, or hormones. The same is true of bedding. Organic livestock feed is often more expensive than other feeds. The use of chemicals, wormers, antibiotics, and similar drugs is eschewed in favor of herbal remedies and strong genetics. Consequently, when life-threatening illness strikes, the organic farmer has fewer tools to treat the illness.

On the plus side, organically raised goats may be healthier and more resistant to disease through natural selection. Although organic goat products typically cost more to produce, they also tend to command higher prices. Goats on strict organic diets provide food that a family can trust.

Sustainable agriculture tries to combine techniques from both organic and conventional systems. Efforts are made to reduce the amounts of antibiotics and chemicals in feed and treatments. However, both antibiotics and chemicals may be employed when necessary. The sustainable goat farm tests for parasites and treats only when worm loads indicate a need. Feeds do not normally contain drug additives but may not contain certified organic ingredients. Goats are put out to pasture whenever possible.

FEEDLOT MANAGEMENT

A feedlot is an operation where animals are raised in such numbers that manure accumulates and prohibits significant groundcover. Municipalities have regulations regarding feedlot management; however, most small family goat farms do not constitute a feedlot.

Check with your state department of agriculture if you believe your operation may need a permit. The main concern with concentrated animal operations is the potential for surface-water and groundwater contamination. Other public concerns include odor, traffic, and aesthetics. Regulations focus on the number of animals on the site and the methods in place for handling waste products. While most goat operations are exempt from feedlot rules, it makes sense to voluntarily follow the best practices in your operation. Running water and drainage should be away from the animal housing and yards. Divert runoff so it does not cross these areas. Be aware of any high concentrations of manure that may leach runoff into bodies of water. A buffer strip of land between the lot and any stream or pond reduces the possibility of contamination.

DEAD ANIMAL DISPOSAL

An unfortunate aspect of raising livestock involves dealing with death. Depending on the size and scope of your operation, mortality may be a rare experience or a regular chore. During kidding season, owners lose kids through premature birth, accident, or illness. Breeding season is hard on bucks, and kidding takes its toll on does.

Municipalities have laws regulating carcass disposal. Rules take into consideration public health and safety, including air or water pollution, spread of disease, nuisance odors, and pest attraction. Check with your state department of health to learn approved disposal methods. State law typically prohibits throwing even small carcasses into home garbage pickup. Burial is common when carcass numbers are small and where climate permits. The carcass should be at least 3 to 6 feet under the surface to discourage scavengers. It should also be at least 5 feet above the water table to prevent groundwater or well contamination.

Cremation, or incineration, is usually regulated. Open burning is prohibited in most locations. Incineration can be accomplished any time of year and has a low biosecurity risk. Disadvantages include the cost of equipment and the smells associated with burning flesh. Consider your neighbors.

Biosecurity risks have decreased the popularity of sending dead livestock to rendering and pet food plants or fur farms.

Caprine management systems range from conventional at one end to organic at the other. A sustainable management system falls somewhere in between. *Shutterstock (top), Jen Brown, Terrapin Acres (bottom)*

Consider what you want from your goats before you select your breeds. *Jen Brown*

Moving carcasses from one farm to another poses risks to both properties. There is no way of knowing what type of illness killed the animals previously transported by the vehicle coming to pick up your dead animals. Many disposal firms won't take goats or sheep for fear of scrapie. Nevertheless, it is worth checking out local fur farms, wild game refuges, or pet-food plants in your area. Make these contacts before they are needed.

Composting as a method of disposal is gaining in popularity. Environmentally friendly, inexpensive, and low-risk in terms of biosecurity, composting should be considered whenever possible. Layer at least 1 foot of sawdust, bedding, or another carbon source below and around the carcass. Properly done, composted carcasses give off few odors and present very little biohazard.

GETTING YOUR GOATS

Before you purchase a goat, it is important to know why you want it and what you are going to do with it. Consider the following questions:

- What do I want from my goats—meat, milk, fiber, or companionship?
- How many animals do I want to take care of?
- What space—both housing and pasture—do I have?
- What goat breeds are available in my area?

If you are expecting to have milk, you definitely don't want a wether (a castrated male). If you don't want to milk twice a day for 305 days a year, that wether may be just what you need. Some breeds, such as the Nubian, are good meat and milk animals. Like dairy cows, most dairy goats can be used for meat as well as milking purposes. Boer goats, traditionally a heavily muscled meat breed, can produce rich milk in quantities sufficient for a family. There are people who milk Pygmies, traditionally a novelty pet, but most choose the more dairy-oriented Nigerian Dwarf if they want a miniature goat to supply family milk. Rare breeds allow you to help preserve unique bloodlines.

EVERYONE IS AN EXPERT

All goat owners have a favorite breed. Don't let the preferences of one breeder determine what is right for you. Visit fairs and shows. Join a goat club, if only for the newsletter. Go to a local conference. Read books and journals. Check out goat-related groups and internet breeder pages. There is a wealth of information out there. The rise of Facebook has made reaching fellow goat owners much easier than when I first became interested in these fascinating creatures. One caveat—get a variety of opinions and be aware that false information is often as easy to get online as facts.

Even with prior planning, your interests may change as you get to know the types of animals available. I started out with a definite preference for the attractive, long-eared Nubian. I also had the idea that I would never own a tiny-eared LaMancha. Yet as my herd grew, sweet, inquisitive LaManchas far outnumbered Nubians in my barn. I have owned, at one time or another, at least one of every recognized dairy breed with the exception of Sables and Saanens. And I have experience with Saanens at Poplar Hill as well as a few that I briefly took care of for a friend. My meat goats have included Boers, Kikos, and several Cashmeres. Currently, my herd is much smaller than when I ran a commercial dairy or meat herd and consists of mostly miniature pet goats of the Nigerian and Pygmy types.

A fellow goat owner told me, "Just when I find the reference for how to handle a certain problem, the goats show me that they haven't read the book!" When I think I can learn no more about goats, it will be time for me to quit raising them.

AUCTIONS

Once you've settled on a goat breed, your next step is to research your options as a buyer. Auctions are one of several possible sources. The auction barn is a decent place to sell goats but rarely a good place to buy one. I will admit this opinion is based on standard local auctions that primarily feature meat animals. The exotic livestock or specialty goat auctions may have animals more suited to breeding and some may even have papers available. Usually, the auction barn is where breeders sell their cull goats—inferior animals that have been removed from the herd. Most of these animals end up as meat. Some nice goats do show up at these sales; usually, they are owned by someone who isn't a breeder and doesn't want to go through the trouble of marketing the animals.

If you do look for goats at an auction, examine the animals before they reach the sale ring. Run your hands over each goat and perform a thorough investigation. A healthy animal is lively and has a shiny coat and clear eyes. Watch for limping, swollen joints, or a misshapen udder. Look at lumps carefully to determine if they are infected abscesses or simply vaccination lumps. Manure from a healthy goat is firm and pelleted. Extreme skittishness may indicate poor temperament.

Those cute kids or that lovely looking goat could have hidden defects. Cull goats may have extra teats, the inability to breed, or chronic illness. Don't be persuaded by a low asking

Don't judge based on looks alone. The LaMancha, with its strange-looking ears, has a lovable personality. *Jen Brown*

IN POINT OF FACT, THERE IS NO CORRECT AND REASONABLE PRICE FOR GOATS. THE PRICE OF A GOAT IS WHATEVER TWO PARTIES AGREE TO, THEN AND THERE; NOTHING ELSE COUNTS. THOSE THINKING OTHERWISE ARE BOTH GULLIBLE AND VULNERABLE; ACCORDINGLY, THEY ARE CERTAIN TO SUFFER THE CONSEQUENCES OF THEIR SHORTCOMINGS, USUALLY SOONER THAN LATER.

—DR. FRANK PINKERTON

price. If the goat is too thin, has a rough coat, or has obvious sores or abscesses, it may not be cheap in the long run. You could pay down the line in veterinary expenses and heartache.

On the other hand, a specialty goat auction, common with meat and fiber goats, is typically a good place to both buy and sell stock. These auctions provide pedigrees, papers, and owner information. Some goat organizations hold auctions for fundraising or breed promotion.

BREEDERS

Most breeders sell not only the goat, but also their knowledge. Some good places to meet breeders are clubs, shows, and exhibitions. To talk to breeders at a fair, the best time to approach them is when the animals are not showing. Shows can get hectic. The owners are frequently busy and lack the time to answer basic questions. If they are busy, ask for a business card or a time you can come back to talk.

PRIVATE PARTIES

Private owners who are getting out of raising goats, or those who keep just a few for family milk and are selling the excess, may have nice animals for sale. These goats are generally better socialized than those from a large commercial herd. You may not receive as much information about bloodlines, but if you don't plan to go into the breeding business yourself, there are real bargains to be found through the newspaper or online advertising such as Craigslist or local farming and goat groups on Facebook.

Once you have the breeder's full attention, ask away. What diseases have been problems in the herd? Have the goats been vaccinated or wormed? What diet and feeding schedule have the goats been on? The answers will guide your management of any animal you buy.

COSTS

Goat prices vary as much as the animals themselves. Unpapered pet goats are usually the cheapest. Like other things in life, you get what you pay for. Research the prices in your area the same way you researched the breed you want. Breeders put ads in goat publications and online; many times they have complete sales lists of their available goats. Read the newspaper classified ads, check online sources, keep an eye on sale barn reports, and go to auctions as an observer to give yourself an idea of the local market.

If you see an ad for $5 newborn bucklings, remember that to make good pets, they will require bottle-feeding, neutering, and disbudding. These animals are being sold because they are excess from a large dairy and do not suit a novice pet buyer's needs.

Registration is an important factor in pricing. Some goat registries are breed specific. Some promote the commercial interests of the breeder; others concentrate their efforts on showing. Some just provide identification and documentation of ownership. Registered goat prices therefore have a wide range. The average is $200 to $400 or more for registered doelings; high-quality papered bucks are often more expensive. Rare breeds and show animals, particularly champions and their offspring, can cost $1,000 or more. That said, registered stock isn't a sure bet. Genetics are based on percentages. Be aware that hidden genes in the parents can manifest in the offspring. Some breeds and lines reproduce consistently. Other breeds just don't reproduce as true to their phenotype.

Beware of cheap kids. They may be cute, but they may not be so affordable in the long run. *Jen Brown*

Unregistered stock can be as nice as—or nicer than—goats that have papers. Remember the axiom "What you see is what you get." Make sure you like the conformation and looks of any goat you are buying. If possible, check out the goat's parents or siblings. Numerous other factors can influence the price of a goat. Some breeders follow a protocol to prevent a devastating disease known as caprine arthritis encephalitis (CAE). Kids raised using these prevention techniques are normally more expensive than kids raised with traditional methods because of the extra work involved. The added value this method brings often offsets the additional expense.

A special comment about "disease-free" herds: Different goat owners have different tolerance for disease. As noted elsewhere in this book, some people maintain herds under conditions of disease management. Other owners prefer disease eradication and will not tolerate certain conditions in their animals. If you plan to start and maintain a disease-free herd, be aware of certain limitations, then decide if you can live within them.

In order to be disease free, learn about tested, contagious diseases and decide which ones you cannot tolerate. Plan on keeping a bio-secure farm, with no untested animals mingling with the clean herd. Take care whether to exhibit at shows or fairs, sanitize pens before moving your animals in, and limit contact with strange animals and their milk as much as possible. Over the years, I have found that the testing, culling, and other things associated with claiming to be disease free was more expensive and onerous than the amount of money I could get for my animals.

If you are being strict about CAE or other diseases, don't trust the word of the breeder who states they are clean unless you are willing to later find out they were mistaken. From untested herds, some sellers offer to test sale goats at the buyer's expense. The only true measure of disease-free herd status is a record of routine (every 6 to 12 months) testing at a certified laboratory.

HANDLING THE TRANSACTION

Once you decide to purchase a goat, how the transaction is handled will affect your satisfaction with the goat and the deal. Follow simple rules of courtesy. If you are buying at an auction or from a person giving up a pet, many of these precautions are unnecessary. However, good breeders offer more information with their goats and generally charge accordingly. These tips will help you get the most out of your purchase:

- Be clear in your expectations. Talk to the breeder. Clarify your expectations about temperament, health status, body conformation, and milk production.
- Settle health expectations before the sale. Educate yourself about the conditions affecting these goats and decide your comfort level. If you want livestock from a disease-free herd, ask about herd health status. If you require blood testing, be prepared to pay the owner up front. Testing can be expensive, but if you feel strongly about diseases, consider the money well spent if the goat shows a positive result and you must reject it. Usually test costs are not refundable if the animal comes up positive.
- Decide your purchase limits before seeing the goat. Goats are lovable. The excitement of getting a new one has clouded many an experienced eye. Clarify your expectations before seeing the live animal. Tell the seller if you are only willing to spend a certain amount. If the price is higher than you can afford, politely thank the breeder and look elsewhere.
- Follow up on your inquiries. It is easy to send an email or message to a farmer asking about goats for sale. Some breeders put a lot of time and effort into their reply. It is only polite—and will give you an advantage if you wish to deal with this breeder again—to express your gratitude for the information. It is appropriate to say, "I guess that isn't what I am looking for right now." Then the seller won't hold back the sale to someone else because you seemed interested and asked first.
- Make an informed decision based on personal observation. The sale is not complete until you say it is. A farm visit allows you to observe the health of the goat and its herd mates. Ask to see the sire and dam, if available. Check the goat for lumps, bumps, parasites, and general physical condition.
- Negotiating is fine, but don't insult the seller. At the farm, discuss your sale

When buying a goat, take care to handle the transaction smoothly. *Shutterstock*

based upon direct observation. If there is something you do not like about the goat, tactfully explain your concern to the seller. It is then okay to offer a lower price if you are still interested. It is also okay to say "thank you for your time" and walk away.

- Pay promptly. If the seller has asked for a deposit, send the funds as promised. Payment in full is expected before the animal is shipped or before you leave the farm, unless a payment plan is prearranged.
- Deposits are nonrefundable. Do not expect your deposit money to be returned if you back out of the sale. The breeder may have turned down other buyers and has cared for the goat while holding her for you. (Deposits are usually returned if the goat is unavailable due to kids not being born or the sudden illness or death of the animal.)

Bucks can be very hard on fences—especially during rut.
Jen Brown

- Be on time. Set farm visits at a convenient time for you and arrive when you say you will. Ask if the seller has any time constraints if you hope to visit longer than the time it takes to pay for and load the animal.
- Pick up the goat in a timely manner. Many farmers have taken a deposit on a goat only to find they are still feeding and housing the animal six months later. If you can't take the goat, contact the breeder to explain the problem. Negotiate boarding if you still want the animal, or release the seller from the deal. If kids have been born to a doe while she stayed in the care of the seller, those kids will be left with the seller unless previous arrangements were made.
- Communicate! Once the goat has reached your farm, whether having been shipped by the seller or picked up by you, communicate with the seller. Let the breeder know the animal arrived safely or if it is showing signs of shipping stress, such as diarrhea or respiratory problems. Follow any treatment recommendations. If the animal is unacceptable for any reason, tell the breeder immediately so you can seek a remedy.

TO REGISTER OR NOT TO REGISTER

If you purchase goats that are qualified for registration but aren't yet registered, you'll have to decide whether to register them. Consider your breed, the reason you are keeping goats, and what you will do with the offspring. Registration papers are usually required if you want to show your goat. They're also useful if you intend to eventually sell the animal or its offspring. Potential buyers appreciate being able to trace parentage and see how an animal fits into its particular breed. If you are a breeder, keeping up with registrations will enable you to attract a wider range of buyers.

HOUSING YOUR GOATS

Goats are tremendous escape artists. A solid fence is your best defense against an inquisitive,

Cattle panel gates can be made stronger by twining two metal pipes or chain-link fence rails through the wire. *Jen Brown*

foraging goat. Avoid improper bracing of the fence with too few posts or placing the posts in soft ground. Bucks will lean on a loose fence panel until it is low enough to walk right over, and such adventures can result in out-of-season kids and damaged shrubbery. Woven wire and chain link are also good options for fencing material. A fence gets more of a workout when located near greenery desirable to the goat. For this reason, these fences are best for larger areas or for younger goats that aren't heavy enough to damage the fence. Light-duty fences with a single "hot" electric wire running alongside discourage leaning and escape. Otherwise, the recommendation for electric fence for goats is seven strands. Goats may be trained to respond to fewer wires. A modified high-tension fence of four strands can work.

One electric fence on the market features woven wire. I have heard secondhand stories of sheep or goats getting tangled in the wire and shocked repeatedly, although many breeders report that it works well for their animals. Proper training of goats to the woven fence can help eliminate this kind of problem.

Hardware clips make for inexpensive latches. *Jen Brown*

Twine is always available around the farm for quick repairs or temporary pen ties. *Jen Brown*

When first shocked by an electric fence, a goat frequently charges through it rather than backing away. Goats may be trained to respect the fence by starting them in a small, temporary enclosure surrounded by electric wire. Once the goat has experienced a shock, it quickly learns to avoid it. The trained goats are then ready to go out into a larger electrified area.

Panel fences come in three main sizes. Hog panels are 3 feet high and have a narrower spacing at the bottom than the top. These fences are nice for kid pens since the kids are too small to climb over them, and the owner can easily reach into the pen to work with the animals.

Cattle panels are my favorite for small yards. They are 50 inches high and hard for adolescent or adult goats to jump. The welded 4-gauge wire is uniformly spaced 6 inches apart except for the lowest two levels, which are 4 inches. The wider spacing can be a problem because small kids can wiggle through the fence.

The most expensive panel—and the best for general use—is the combo panel. This variety has the 4-foot height of the cattle panel and the dense bottom spacing of the hog panel. Depending on the number of goats testing the fence and how badly any particular goat wants to escape, gate security can be a real chore. I'm still trying to find the ideal gate clip that allows me to access the pen easily. One-hand operation is optimal. I usually use hardware clips, which are strong enough to handle fence pressure and don't break too frequently.

Our state fair goat pens used to have latches that some goats opened with ease. One champion LaMancha had to be put in a specially secured pen at the fair since she routinely took herself and her pen mates for walks down the center aisle after opening the latch!

Bungee cords work for short-term latches. The goats like to use their mouths on them, but they hold for limited periods. On the downside, the elastic part of the cord ages quickly, and certain goats enjoy chewing them apart. In a pinch, baling twine is always cheap and easily replaceable. Just remember, some goats like to work the knots out with their agile mouths!

TIPS FOR A GOOD ELECTRIC FENCE

1. **ATTACH STAPLES CAREFULLY.** A staple hammered through insulated fencing will cause a hidden short.
2. **AVOID BOTTOM WIRES GROUNDING OUT.** Wires in contact with wet vegetation or snow greatly reduce the charge. Allow for bottom wires to be shut down in wet snow or grass.
3. **BEWARE OF OLD FENCES.** Running your electric fence along an existing fence line is tempting; however, the old wire has a way of contacting and shorting out the new electric wire.
4. **FACE THE SUN.** If using a solar panel charger, face the charger toward the sun. Inadequate sunlight means a poorly charged fence.
5. **USE GOOD INSULATORS.** Plastic and sunlight don't always mix well. Poor quality plastic will turn white or clear with extended exposure, while good insulators are treated to avoid breaking down.
6. **KEEP WIRES APART.** Allow at least 5 inches between wires and be sure that any spacers or strainers are well separated and cannot cross each other.
7. **DON'T MIX METALS.** Combining copper and steel weakens the fence as electrolysis corrodes the wires.

IF A FENCE WILL HOLD WATER, IT WILL HOLD A GOAT.

—PROVERB

Heavy-gauge wire is a longer-lasting alternative to twine and less costly than commercial fence clips for tying fencing of many types to t-posts. *Jen Brown*

One or two electric wires may be offset from a nonelectric fence to improve the barrier. *Terrapin Acres*

Tension for electric fencing is crucial. These are tensions bars for a gate on a high-tensile or "New Zealand" type of electric fence. *Terrapin Acres*

8. **PROVIDE PROPER GROUNDING.** Good grounding requires several 6- to 8-foot well-attached galvanized rods and a complete circuit from fence to ground.
9. **REPAIR DAMAGE PROMPTLY.** Kinked or flattened wire breaks easily. Damaged sections of fence should be spliced with a fence splicer or a hand-tied square knot.
10. **SPACE POSTS AND TIES PROPERLY.** Electric fencing should be elastic enough to spring back when hit by animals. If the electric wire is too tight or the fence is built with too many posts, the fence, insulators, or posts can break.
11. **THINK BIGGER.** If you are tempted to skimp, think again. A charger with no "zap" won't impress the livestock, and a thin wire will carry a thin charge. Even with a large charger and heavier wire, the electric fence is still the least expensive large-area perimeter.
12. **USE A VOLTMETER.** Check the fence with a voltmeter (instead of your hand).

SHELTER

When protected from wind and wet, goats can withstand extreme temperatures. The most basic shelter that meets these criteria works for

Some goats have humble abodes (top); others share large, open barns (bottom). *Jen Brown*

These bucks live in a converted grain bin. *Jen Brown, Terrapin Acres*

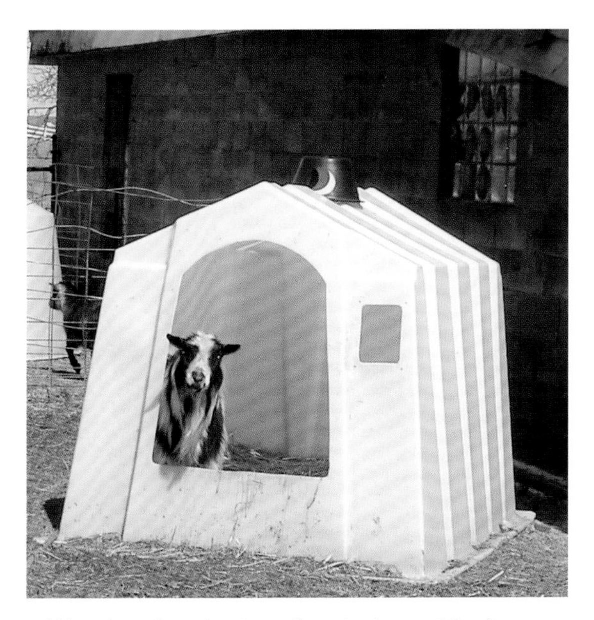

Calf huts have the advantage of coming in a variety of styles and sizes while being readily portable. Unfortunately, weather, animal activity, and time all contribute to breakage. *Carol Amundson, Terrapin Acres*

Goats and other small ruminants are generally hardy animals that can withstand very cold temperatures, so long as they also have shelter from wind and wet. *Terrapin Acres*

goats. Any variations are for the comfort and pleasure of goat owners. My goats have lived in sheds, calf huts, a barn with dirt floors, a barn with concrete floors, a converted grain bin, and dog houses. Imaginative owners have housed their animals in culverts, travel trailers, chicken coops, and wherever the darn things decided to make a bed!

The recommended living space is 16 square feet per full-size goat. More crowded conditions are feasible. Goats like each other's company and congregate by inclination. Allowing more

A useful homemade flytrap can be made with an old milk jug filled with water and a fly attractant, and then capped with a trap lid. Flies enter and eventually drown in the liquid. Chickens make more attractive fly catchers, although their droppings can be a nuisance. *Carol Amundson, Terrapin Acres*

space, however, spreads out their waste and helps keep the bedding cleaner. Crowding also increases the risk of disease, especially parasites.

Calf huts are easy to clean, moveable for cleanup, and relatively inexpensive. Drawbacks to the least expensive models include cracking and quick breakdown of the polymer plastic or fiberglass from sun exposure. Huts can be difficult for owners to maneuver within when catching their goats.

Barns and sheds are advantageous. The accessibility of a pen in a barn beats the hut hands down, especially when the goats are kidding or sick. My own barn has a concrete floor that was there when we bought the farm. This material is hard to clean and requires a deep bed of straw or other bedding for comfort. A dirt floor is better for absorbency and comfort for the animal's feet.

Poplar Hill Dairy is particularly well designed. The animal pens have dirt floors, while the aisles for human traffic are concrete. The concrete level is higher than the pen level, which allows bedding to build up in the pens without spilling into the aisles. Along the aisles are fences that double as feeders. Hay can be placed right on the concrete for the goats to reach. Waste hay and dust can be swept into the pens for easy cleanup. Doors on the ends of the dirt pens allow access for cleaning up manure.

FLIES

Where there is livestock, there are flies. Under optimal conditions, fly eggs mature into adults in as little as three days. Your farm may experience as many as ten to twelve generations of flies in one summer. Flies make the animals and caretakers uncomfortable—and they spread disease. Milk production and weight gain are both restricted when animals are constantly fighting insect pests. In short, they are a menace!

Do not give the flies a place to breed. Keep your facilities clean and dry. The ammonia in animal waste is particularly attractive to flies. Spreading agricultural lime on floors and in bedding discourages flies. Spiders, bats, and barn swallows should be welcomed, as they help control the fly population.

Flytraps may be purchased or homemade. Bags or jugs of water containing an attractant are placed as lures. Place them where they will not be knocked over or grabbed by the goats. Some attractants, such as sugar, leave a sticky mess when spilled, and pheromone traps often smell putrid.

Sticky tape strips or streamers may be hung in the barn. They trap large numbers of flies but also catch beneficial insects. Hang fly tape away from the flight path of birds and bats.

Poultry, such as chickens, ducks, and guinea fowl, eat a tremendous number of bothersome insects, including flies and ticks. The free-range chickens in my goat and cattle yards spend hours scratching up animal waste, eating fly larvae before they even have a chance to hatch. If you plan on getting certification for your dairy, be aware that many states have dairy laws ruling that you can't have poultry in the dairy yard. This is a shame, because I have had much better luck reducing the fly population with free-range chickens than with any other control method. In addition, open housing and plenty of room without overcrowding your animals is one of the best methods to reduce insect pests.

I prefer to use natural fly-control methods. However, hiring a pest control company is easy and certainly keeps flies at bay. You can also use your own chemical application. Chemical risks include contamination of food products, destruction of beneficial insects and wildlife, and development of resistance in the flies you are trying to control.

FEEDING YOUR GOATS

Myths aside, goats are picky eaters. They may taste many things but frequently do not eat what they examine. Water and hay contaminated by feces remain untouched unless the goat is driven by desperation. Some goats are suspicious of new foods and must be almost starving before they will try unfamiliar products. Other goats will reject the offering after tasting or feeling the texture. Proper feeding ensures that goats live longer, are more productive, and have fewer health issues overall. The largest annual operating expense for livestock producers is normally feed-related.

The feeding of ruminants is a science in itself. One excellent reference about livestock feeding is the website Feedipedia, an "open-access information system on animal feed resources that provides information on nature, occurrence, chemical composition, nutritional value and safe use of nearly 1,400 worldwide livestock feeds." New feeds or a rapid increase of protein or fat can throw off the balance of the delicate chemical processes in the rumen and lead to potentially fatal conditions, such as diarrhea, bloat, or enterotoxemia. Changes in diet should be administered slowly and monitored for signs of trouble. The basic elements of a goat's diet include water, hay or roughage, grain (also known as concentrates), and supplements.

> THE GOAT ATE THINGS. HE ATE CANS AND HE ATE CANES. HE ATE PANS AND HE ATE PANES. HE EVEN ATE CAPES AND CAPS.
>
> —SIEGFRIED ENGELMANN AND ELAINE C. BRUNER, *THE PET GOAT*

Goats are playful and will explore any new props you put in their yard. *Jen Brown*

WATER

Water makes up more than 60 percent of a goat's soft tissues. Goat milk is 87 percent water. All of this fluid needs to be replaced daily. When they are under stress, especially when traveling to a show, goats sometimes refuse to drink water. Try adding vinegar, electrolyte solution, or even Kool-Aid to entice them to drink and stay hydrated.

Have clean water available to the goats at all times. Place water buckets and troughs where the animals can drink easily but also where they cannot foul the water with waste, such as on the other side of the fence. The parasites, bacteria, and viruses that live in contaminated water can lead to lower milk production and sickness.

During warmer weather, it can be difficult to keep ahead of the growth of algae in water tanks or buckets. Rinsing with a mild bleach solution as necessary will reduce the green goo from accumulating in tanks or buckets. Some farmers keep a few goldfish in the larger water tanks to eat insects and algae. Care must be taken to keep the tank full so the fish don't die.

In a kidding pen, water buckets should be placed in a spot where there is no danger of kids being born into the bucket. Newborns, as well as lively older kids, may fall into a bucket and not be able to get out. Drowning is a tragic loss of a healthy kid. I never saw this problem with my full-sized goats, but unfortunately lost a miniature breed kid in a standard 5-gallon water bucket.

Automatic waterers can be installed. A number of commercial models work with either troughs or floats and allow animals a continuous supply of water. Another option is the Lixit or nipple valve. This simple device is the same as the nipple on a water bottle used in a small-animal cage. The larger version used for goats is mounted in a plumbing line or a container beside the pen.

When temperatures drop below freezing, watering becomes a major challenge! In Minnesota, for example, temperatures stay below freezing for weeks. Goats eat snow and lick ice but not enough to meet their water needs. Heating or insulating the water container is necessary. Heated water systems are the most common solution for winter watering. Floating heaters are placed in tanks or buckets. Nonmetal containers melt or burn when contacted by the metal heating element in these types of heaters, so keep the water level high and use heaters with guards. Keep electric cords away from curious and playful goats.

Keep clean water available to the goats at all times. Water needs to be changed regularly. If you keep other livestock, particularly waterfowl, they will also use the goats' water. One ounce of bleach per 15 gallons of water can be added to reduce the growth of microorganisms. Careful! Too much bleach can upset the rumen organisms.
Jen Brown

Stored hay loses nutrient value. The first things to drop are vitamins A, D, and E. *Jen Brown*

Buckets with a heating element built into their structure are a great investment. The cord is hidden within the walls of the container. The water is easily accessible to the goats, as there are no floating heaters to get in their way. Larger automatic watering systems are available in heated models with thermostats that regulate heat during freezing temperatures.

HAY AND OTHER ROUGHAGE

Good-quality hay is important for caprine nutrition. The average goat eats 4.5 pounds of hay a day per 100 pounds of body weight. Hay provides protein and vitamins, as well as fiber to keep the rumen working properly. Fiber provides energy, particularly in milkers, young goats, and does in late pregnancy. The younger and fresher the hay, the better the quality. Good hay is expensive, but the better the hay, the less grain and supplemental feeding you'll need. High-quality hay is especially important in the dairy barn. Good quality hay can make immediate improvements in the amount and quality of milk produced.

You may find less expensive hay sold in your area advertised "for cattle or goats." Often this is older hay or hay that is very stemmy. The goats will waste large amounts of poor hay making it less of a bargain. Goats don't utilize the cell wall of plants as efficiently as cattle. Low-quality hay that has mostly stem and few leaves or seeds will not be a good food for goats. The amount of feed needed by meat goats is said to be twice that of cattle relative to their size.

Finding Hay

If you don't grow sufficient forage on your property to feed your goats, you'll need to purchase it from a reliable source. For the pet goat owner, a few bales at a time from the local feed store or a neighboring farm are enough. Large operations may have hay shipped to the farm by the ton. Different parts of the country have different types of hay. In the Midwest, alfalfa, grass, and clover are common. Buy the best hay that you can afford. Store it properly, preferably under cover and off the ground. It may seem like a bargain to buy hay that is one or two years old. Unfortunately, older hay is better for roughage or bedding and may leave your goats deficient in actual feed value.

Stored hay loses nutrient value. A field dotted with large round bales of hay may look beautifully pastoral, but this method of storage causes drying and loss of digestibility. A round hay bale stored outside may lose 15 to 25 percent of its feed value within the year. It is better to store hay in the protection of a barn. Depending on the moisture content, hay stored under cover can retain most of its nutritional value for up to a year. Putting up hay that is too wet leads to mold and microorganisms, which consume nutrients and release toxins that can cause serious illness. Never feed your goats moldy hay! In a worst-case scenario, overly wet hay stored in a building can spontaneously combust. The resulting barn fire would devastate any operation.

Green, leafy hay is high in vitamin A, one of the vitamins necessary for healthy goats. They also get this nutrient from corn or good pasture. *Carol Amundson, Terrapin Acres*

Tree branches are a treat for goats that don't have browse in their pasture. *Jen Brown*

PASTURE OR RANGE

Pasture and browse can supplement hay. Don't try to maintain a goat herd strictly on pasture. Young pasture that is rapidly growing may contain too much moisture for the goats to eat enough to meet their protein and nutrient needs. Drought conditions change the composition of plants and can also concentrate toxins. Allowing pasture goats to have hay at least once a day ensures that they are properly fed. Before allowing goats out to a new pasture it is helpful to give them hay or dry forage so they don't get upset digestion from gorging on rich new greens on an empty stomach.

Goats are natural browsers. They will eat selectively and appreciate the opportunity to forage among woody plants and grasses. This makes them excellent for clearing areas that can otherwise be difficult to cut. I have enjoyed watching a goat very carefully select just the blossom from a thistle, while avoiding the worst of the thorns. Be aware that there are poisonous plants. Goats should not eat yew, bracken fern, or hemlock, among other things. When giving your goats prunings from the garden, avoid rhododendron clippings or prunings from cherry, apricot, or peach trees, which can contain cyanide when wilted.

A goat's natural eating style is to wander about and pick and choose from a variety of browse. Goats are useful for cleaning unwanted brush. Brush has lots of fiber, plus vitamins. Be aware that even the bushiest area will not survive intensive grazing for more than two seasons.

SILAGE

Special warning should be made about silage, which is generally not recommended for feeding goats. The risk of mold, bacteria, or even excess nitrates is high. Goats are particularly susceptible to mold and other spoilage. Listeriosis and botulism have been associated with eating spoiled silage. Goat owners have learned this lesson after tragically losing a large portion of their herd to illness and death following the feeding of seemingly good quality silage. I recommend finding a goat mentor with experience in this type of feed if you choose to use silage.

BALEAGE

Baleage is similar to silage. It is chopped hay that is harvested at a higher moisture than traditional dry hay and wrapped in plastic wrapper. While silage usually has a moisture content above 65 percent, baleage (also known as haylage) has a moisture content of 40 to 60 percent. Because of the higher moisture content, more baleage or silage is fed per animal by pound than hay.

COMMERCIAL PACKAGED FORAGE

There are packaged forages available for goat owners looking for a more standardized feed, especially in areas without ready access to alfalfa or similar-quality hay. One of the most common commercial forages is Chaffhaye, a fresh, chopped, bagged alfalfa-based feed

During the summer, goats that are not growing rapidly or milking heavily may get all their nutrition from good pasture. Pregnant goats, does in heavy milk production, and adolescents usually need supplements of other high-energy feed. *Jen Brown*

Branches placed in the goat pen quickly get stripped of leaves and bark. *Jen Brown*

Bag-style horse feeders serve as temporary feeders at a fair for some young goats. *Carol Amundson*

with a guaranteed level of nutrition, a non-GMO certification, and a 16-month shelf life. Chaffhaye is processed to retain its natural plant juice, preserving the natural probiotics of the feed. It is advertised as "Premium Pasture in a Bag." Goat owners who use this often report improved production and health. Chaffhaye is a premium product and is often notably more expensive than local hay. Some breeders use it as a supplement to hay or alfalfa pellets, reserving Chaffhaye for heavy milkers and pregnant or older goats.

HAY FEEDERS

Hay feeders create unique puzzles. The ideal feeder holds hay up for goats to reach comfortably while preventing scraps from falling out of the feeder and being wasted. The perfect feeder also allows goats access to hay without trapping their head or horns. The feeder is tight enough to keep smaller goats from climbing in and spoiling the hay with their dirty hooves or waste. The feeder capacity is large enough to hold a half-day's worth of fodder. Naturally, this mythical feeder does not exist. You'll need to find the compromise that works best in your situation.

For adjacent pens, feeders that hang on both sides of the fence can be used. Some of these are good only as temporary feeders because they get easily banged around and battered by the goats. *Carol Amundson*

A fence-line feeder is a common and easy hay setup. *Terrapin Acres*

Bag and Net Feeders

Typically used for feeding horses, a bag (or net) feeder hangs in the pen either on the fence or wall or suspended from the ceiling. The feeder must be close enough to the ground to allow the goat to get at the hay but high enough that the goat does not get tangled in the net. I have read about goat owners who swear by this feeding method. I have never used it, as this design seems to create the risk of legs being trapped and then broken by the goat's ensuing struggles. Worse still, the goat's head could be caught, causing it to strangle to death.

Hanging Feeders

One type of hanging feeder is a simple, inexpensive feeder that slips onto a fence. Portable and easy to handle, it holds one or two flakes of hay from a small, square bale. Some of these feeders are flimsy and do not hold up well to the rough treatment that goats can dish out. The wires bend and break off. Once this occurs, the feeder becomes useless because it wastes more forage than it feeds. If you just have a couple of pet goats, though, a hanging

Using fence panel feeders can be a danger to horned goats, who get their horns stuck in the fence. *Terrapin Acres*

The wire in this feeder is sturdy and will hold up to hard use. *Carol Amundson*

A sneaky Nigerian kid slips through the fence. With any feeding setup, you need to figure out how to give larger goats access to the food while preventing kids and smaller goats from slipping through and fouling it. *Terrapin Acres*

feeder is feasible. It is also a good option for a temporary isolation pen in your barn or a pen at the fair.

Fence-Line Feeders

A fence-line feeder uses the fence as a barrier while providing feed at the same time. Cattle panel fencing allows goats to stick the muzzle or head through the wire to get at the hay on the other side. For larger goats or those with horns, some of the holes should be enlarged so the goats may maneuver in and out of the feeder. Be aware, however, that every enlarged hole becomes a possible escape route for kids and smaller goats.

Homemade feeders from wood or a combination of wood and fencing are inexpensive and useful and limited only by the breeder's imagination. *Jen Brown, Poplar Hill*

Mangers and Keyhole Feeders

A manger is a wooden trough or open box that allows goats easy access to hay without much waste. A keyhole feeder is a variation of the manger with small holes that individual goats can fit the head through to eat. Featured in many beginner goat books, this design has both pros and cons. One negative is that a goat cannot see when her head is in the keyhole. This prevents the animal from noticing a more aggressive goat coming from behind to slam her away from the food. She may also have difficulty getting her head out of the keyhole and can injure herself if she tries to exit quickly.

GRAIN OR CONCENTRATES

Every goat needs some grain during its life. The amounts and types of grain that should be fed depends on the type of goat and the quality of the available forage. Kids and growing goats, working goats, lactating does, and those in late pregnancy should all receive grain regularly. Many pet goats and non-lactating does receive too much grain and become fat as a result. Goats generally eat as much grain as they can get. Never change a goat's feed abruptly and always watch for bloat or other digestive issues when you switch. Changes in the type or amount of grain should be made gradually to avoid illness or even death.

Feed recipes abound. Commercial goat mixes may be purchased. Individual feed mills may sell their own mix. Some breeders feed goats a 16-percent "sweet mix" made for cattle and horses. When getting goats, ask local breeders what they feed. You should get a wide variety of answers based on the size of their herd, the crops they grow, or the type of feed-stuffs available to them—people who live near or work in bakeries, potato chip factories, and other food-related businesses will use stale or leftover products as a small percentage of their goat feed concentrates.

I have fed a personal recipe that had twenty different ingredients, including sunflower seeds and alfalfa pellets, mixed for me by my local feed mill. I have also fed pelleted "complete" feed premade by the mill. Breeders in different parts of the country will find that concentrated feeds, like hay, vary by their locale. If sugar beets grow in your area, beet pulp may be used. In cotton country, some feed contains cottonseed meal. My local goat dairy feeds a ground feed that looks like coarse, powdery cornmeal. This feed consists of ground corn on the cob, soybeans, salt, and mineral. The first time my goats saw it, they were very unsure. Within a short while, they, like the Poplar Hill herd, were eating it in big mouthfuls without hesitation.

Currently, my goats eat wet brewers' grain, the leftover solids from the craft beer industry. We give this to our goats free choice as it is a high-fiber feed, low in carbohydrate, high in protein, and very digestible. It seems to cause very little bloat or other digestive issues. I suspect that there is a better trace mineral content in this feed, as I have seen improved coats and health in goats switched to it. One of my older does with cataracts had visible improvement in her condition after about a year on this feed.

A bolt cutter is indispensable. Keep it near the barn for easy access when a goat is stuck in the feeder or fence. *Jen Brown*

There is a long history of using brewing by-product as animal feed. Farms and monasteries in Europe and the Americas brewed their own beer and then fed the mash to their livestock. Barley is the main grain used for brewing, but beer is also made from wheat, corn, rice, sorghum, and millet. The content of brewers' or distillers' grain will vary with the type of beer or alcohol produced. It is a highly variable byproduct and its nutritional value depends on the grain used, temperature at processing, fermentation, and so on. If you have a local brewery or distillery, they will often be very happy for you to take the grain for nothing. In other areas, brewery or distillery byproducts can be purchased wet or dried when there is active industry to support it.

Handling wet brewers' grain is labor intensive. It is wet and heavy—75 to 80 percent fluids. Luckily, our current brewery is very good at removing the liquid. Our breweries have used various types of barrels to hold the grain. The weight has been between 250 to 350 pounds per 50-gallon barrel filled ½ to ¾. Depending on weather, the grain can be highly perishable. If we allow the top layer of grain dry to a crust, even in hot weather, the grain maintains its palatability up to a week—once the top few inches are thrown aside for compost or chickens. In winter, the grain freezes and can be difficult to shovel out of barrels. On the other hand, the goats seem to enjoy chewing on giant grain popsicles.

Whatever concentrates you choose to feed your goats, keep an open mind and be aware of your local options. For someone wanting simplicity of feeding for a few goats, I recommend using your local mill's general goat feed or a high-quality commercial feed. If you are interested in going in-depth or have a larger commercial herd where costs directly affect the bottom line, explore your options further. Some feed mills will give you the assistance of a feed specialist, and the science behind feeding a ruminant fills pages on the internet.

GRAIN FEEDERS

There is a wide variety of grain feeders on the market. When deciding on a feeder, the goat keeper must again weigh convenience, cost, and the goats' tendency to climb and play. Consider the amount of space needed for each goat to reach the grain at feeding time. More aggressive animals often prevent meeker herd mates from eating. Owners should separate feeders and allow enough head space for all the animals to have a fair chance.

Creating a fence-line feeder can be as simple as dropping the grain on the ground or concrete outside the fence and allowing goats to eat it by sticking their heads through. This can work in an inside pen that is protected from weather. On the other hand, uneaten grain can create quite a mess outside on dirt or grass. It can also lead to increased parasite loads due to ground contamination with organisms.

The Milk Stand or Stanchion

Individual feeding stands are highly desirable for a small goat herd. You can control the amount of food each goat eats by locking the goats into individual stanchions or tying them by a feeder. At feeding time, portion each goat's ration and add any special wormers or supplements that an individual goat may need. This method is impractical in larger herds except dairies, where milking does are fed on the milk line. Like people, different goats have different metabolic rates. Goats that have lower feed requirements are known as "easy keepers." An overweight milker needs additional grain to milk well; however, the extra grain aggravates her weight problem. One trick is to feed her beet pulp soaked in warm water. This fills her up and promotes milk production while not giving as many calories as concentrated feed.

SUPPLEMENTS

Supplementation is an evolving science, with even the experts constantly learning more about which vitamins and minerals are best for goats. Like hay and grain, supplementation needs vary by type of goat, season, region of the country, and quality of forage and grain being fed. Fed free-choice or added to a grain mix, these supplements come in a variety of forms.

Goats will step in their feed when they can—causing contamination from feces and possibly spreading intestinal diseases. Keep feeders off the ground when possible. *Carol Amundson*

In a slant bar feeder, even goats with big horns can eat, yet be separated from their herd mates to prevent fighting. Plan on at least 1 foot of feeder space per goat. *Carol Amundson*

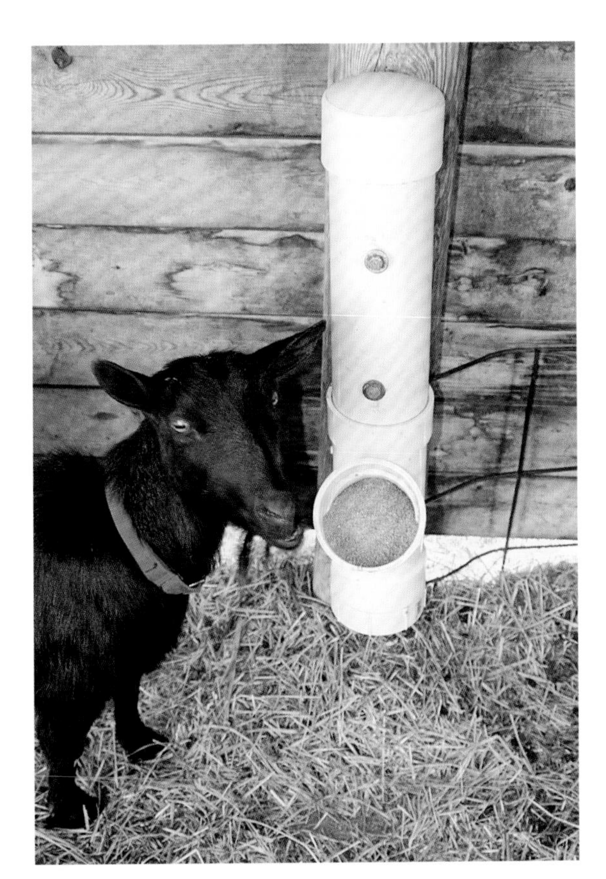

A nice version of a loose mineral feeder may be made using PVC plumbing supplies. *Barb O'Meehan*

Loose Minerals

Loose minerals are best placed in free-choice feeders. General goat mineral formulations are available, as are specific formulations for meat herds and dairy herds. A mineral called Buck Power keeps bucks in prime breeding condition. If goat mineral blends are not available, formulations for cattle or horses can be used.

Blocks

Blocks are formed of compressed quantities of salt, sulfur, cobalt, or a mix. Blocks are durable and weather resistant, although goats may cause wear and tear by playing "king of the hill" on them. Some experts claim that goats can't get sufficient mineral from a block and prefer loose mineral for this reason. There is considerable difference of opinion about how to supplement and in what form that supplementation should take.

Wet blocks are feed supplements bound together in a molasses base. They are beneficial for pregnant does in late fall and winter; the closer certain goats come to their due date, the more they use their block for energy and protein. Some goats become very messy as they spend their time licking at the block and their faces become sticky with the molasses.

COPPER NEEDS IN GOATS

A generic sheep-and-goat feed used to be recommended for both animals, but that has changed. Studies have shown that sheep and goats have different copper needs. Sheep are sensitive to copper, and too much in their diet is harmful, whereas goats require more copper in their diet for proper immune function. Copper deficiency can lead to a whole host of health issues, including coat problems, lameness, weight loss, diarrhea, poor milk production, pneumonia, abortions, and birth defects.

By the same token, it was once believed that the high copper content in horse feed was toxic to goats. It is now known that the copper needs of horses and goats are similar, and horse feed is acceptable for goats. (A word of caution: Horse feed is also high in fat, and too much can lead to obesity or urinary calculi in goats.)

Most pastures and hay types, particularly alfalfa, do not provide enough copper to meet the needs of goats. That said, do not automatically give your goats a copper supplement! Too much of any mineral can be toxic, causing health problems or even death. The best way to determine how much copper your goats are getting is by testing liver samples from dead animals. If your goats are copper deficient, give them a yearly supplement in the form of a bolus. Mineral supplements should be administered in the fall, two to four weeks before breeding.

Baking Soda

Baking soda deserves special mention. Giving your goats free-choice soda in their mineral feeder buffers the rumen and helps prevent stomach problems.

Ammonium Chloride

Male goats, particularly wethers, are prone to a condition called urinary calculi in which stones form in the urinary tract from a concentration of salts. Adding ammonium chloride to the daily feed at the rate of 1 teaspoon per 150 pounds of body weight acidifies the urine and prevents stones from forming. Urinary calculi can be both painful and life threatening, so I consider this a necessary additive.

Probiotics

The addition of yeast or other probiotic mixtures can help the rumen function and improve feed utilization. For some goat owners, this is a routine addition to a feed mix. Others use it only for goats under stress or in ill health. Dry and wet versions are available commercially.

Boluses

Some supplements are given as a bolus, or intravenous injection, on a timed schedule to cover deficiencies in certain nutrients, including selenium and copper. Selenium supplementation, most often called Bo-Se, is recommended for goats in selenium-deficient regions of the country. Copper is frequently supplemented by giving boluses of fine copper wires.

ELDER CARE

Older goats have special needs. Goats are considered aged at five years. By eight years of age, most goats exhibit signs of aging. Teeth wear down or fall out. Additional grain or liquid supplements may be required to improve overall condition. Give senior goats easy access to food and water. Arthritis or stiff joints may cause pain that prevents them from eating enough food or drinking enough water. When older goats are housed with younger animals, competition is fierce, so be sure that elders get what they need to be healthy.

Your herd may contain goats of all ages and sizes competing for food. It is important to have a good feeding setup, where smaller and weaker animals can get adequate feed, unlike this one, which is inadequate for the number of goats shown. *Shutterstock*

3 BREEDING YOUR GOATS

THE MORE THE BILLY GOAT STINKS, THE MORE THE NANNY GOAT LOVES HIM.

—BELGIAN PROVERB

Goats are prolific and hardy. Expect healthy yearlings to produce one to two kids per year and older does to produce two to three. Five to seven kids from a single dam in a year is an unusual occurrence. The Nubian, Boer, and Nigerian are known for multiple births.

Goats with obvious defects should be culled from your herd. Never sell a cull animal to an unsuspecting buyer without full disclosure. In the case of serious defects, the animal should be put down or not bred and a buck sterilized if being kept as a pet.

Some genetic faults are more tolerable than others, depending on the purpose of the animal. Dairy farmers do not want extra teats on a milker. In meat herds, extra teats may be disregarded. Show herds require strict culling for superficial traits that would not bother a pet goat owner.

It is useful to understand basic genetics before starting a breeding program. Each parent goat randomly provides one set of genes to the kid. Think of genetic inheritance like the flip of a coin. The kid inherits a pair of genes that may be alike or different. Identical genes are known as homozygous; different genes are called heterozygous.

Blood or hair testing has become common for the genetic defect G6-Sulfatase Deficiency in Nubians. There is increasing evidence that certain gene locations are associated with classical scrapie susceptibility and resistance in goats. The scrapie test is available but not approved at this time. The G6S test is recommended when breeding Nubians and is discussed in the disease section of this book. I am certain that over time there will be additional genetic testing as science improves our knowledge of goats. After all, dog breeders can now get entire panels before breeding.

When a kid inherits two heterozygous genes from the parents, the dominant gene shows in the kid's appearance. The appearance of an animal is known as its phenotype. Purebred animals have been bred for certain phenotypes over many generations, giving the LaMancha its tiny ears, the Pygmy its size and body structure, and the Angora its fleece. The dominance of a trait is often apparent in crossbred animals. A LaMancha usually produces offspring with smaller ears, for example, and the muscle development and color of the Boer is typically dominant.

A recessive gene is one that hides in the genetic background. Recessive traits only become visible in the phenotype when the kid inherits two recessive genes. A dam and sire may both lack a visible trait but pass that trait to their offspring through recessive genes. Colored Saanens, now known as Sables, are an example of a recessive gene unexpectedly springing up.

Improvement is the goal of the breeder. But improvement is also in the eye of the beholder. One person may be looking for increased milk production, another for more mohair, and yet a third for better attachment of the udder. Some breeders intentionally cross breeds to create new types of goats or improve the vigor of their animals. Meat breeders increase growth rates and survivability through crossbreeding. Each breeding will show improvement and regression of multiple traits.

THE BUCK

There's an old saying among goat keepers: "The buck is half your herd." The buck's impact on a herd is typically greater than that of a single doe, since he may sire substantially more kids than she can produce in a lifetime. Buck selection is both a science and an art. The first rule is to breed what you like to see. Look for a buck that excels in the traits you value and shares the good traits your does already possess.

Opposite: Because he will invariably produce more offspring than any single doe, this buck is "half the herd." *Shutterstock, Kelly Whalley*

Research available bucks to ensure that they have the proper genetics to improve your herd. Do not look only at the buck; some really ugly bucks have sired fabulous kids. Similarly, some fine-looking bucks are known to "throw" extra teats or an underbite or overbite. Check the buck's dam, sisters, and kids.

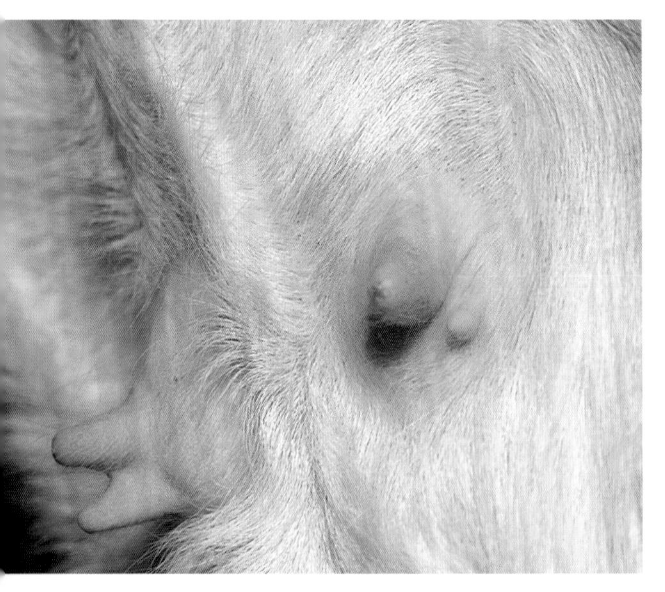

Depending on what you want out of your goat herd, extra teats may be grounds for culling. It's not as important in a meat herd but a negative trait in a dairy herd. *Jen Brown*

WHEN TO BREED

The best conception rates and the healthiest pregnancies occur when goats are in good condition. High-quality feed, worming, and proper mineral levels are important. Too much of a good thing can reduce conception rates, though, so make sure your goats do not go into the breeding season too fat. Some breeders practice flushing, which entails gradually increasing the quality and amount of feed in the weeks prior to breeding. Flushing can help the doe ovulate more eggs. Kidding during the hottest and coldest months is hard on the farmer and the goats.

HOW TO BREED

There are several options for breeding, each with its own advantages and disadvantages. Factors to consider include the number of goats owned, proximity to the buck's residence, type of commercial operation (if any), and whether or not the breeder needs to know exact breeding dates. You may find yourself using a combination of methods.

A buck in rut follows a doe. *Terrapin Acres*

An excited buck checks a doe. *Jen Brown*

A doe teases a buck from the other side of the fence. *Jen Brown*

Caprine gestation lasts 150 days. Think before putting your doe in with a buck simply because she is in heat. Look at a calendar. Some goat keepers plan matings around show season or market factors.

PEN OR PASTURE BREEDING

The easiest way to breed is to put the buck and does in a pen or pasture together and let nature take its course. A buck can service more than two dozen does with a reasonable conception rate. Forty-five days covers two breeding cycles, and most does will conceive in this time if it is breeding season. After this, as the saying goes, familiarity breeds contempt. When does and a buck are kept together for more than a couple of months, they often lose interest in each other. Separating the sexes occasionally might help. The opposite is the "buck effect," in which a newly introduced buck causes does to cycle into heat.

RADDLE HARNESS

A variation of pasture breeding is the raddle harness, which is placed on the buck and equipped with a special crayon on the chest. (Traditionally, to raddle something is to mark it with red ocher.) An alternative is to spread a colored paste on the buck, with no harness; the paste must be applied more frequently than a harness crayon needs to be replaced. When the buck has a solid mount on a doe, the marker colors the doe's rump. The more mountings, the more color. The breeder changes the color of the marker every seventeen to twenty-one

MATING SYSTEMS

TYPES OF BREEDING	DEFINITION	ADVANTAGES	LIMITATIONS
Outcrossing	Breeding two animals of the same breed but with no common ancestors for the past four to six generations	Brings in strong dominant traits Creates hybrid vigor, including longevity, better growth, and improved reproduction Hides bad traits by keeping them recessive	Greater variability in offspring Improvement is based on selection and availability of superior genetics
Line Breeding	Breeding two animals with a relationship in pedigree for a low level of selective inbreeding	Fewer risks than continued inbreeding Greater uniformity of type Helps locks in strong traits Increased prepotency	Increased chance of recessive defects Slow improvement of line, especially if it is mediocre
Inbreeding	Breeding two animals that are directly related, such as mother, father, or full sibling	Carrier animals can be identified and culled Helps detect inferior genetics Increased homogeneity of type in offspring Increased prepotency	Creates inbreeding depression, resulting in loss of size and fitness Higher risks of kids with defects May reduce available genetics in future relationships. Rigid culling required

days until no further marks are seen on the does. Tracking the color on the doe's rump on the calendar identifies probable breeding dates. In this way, raddle marking offers a method of tracking breedings without the need for daily monitoring. This method is used most often in meat and fiber herds kept on range.

HAND BREEDING

Taking the doe to the buck and monitoring the event can be done by keeping the buck on a leadline or by using a small pen for the buck and doe to visit for a short time. If the doe is in heat, breeding requires five to ten minutes for the buck to mount, ejaculate, and repeat the process. You'll know the breeding has occurred when the doe arches her back sharply and the buck throws his head back, then falls or staggers backwards. Once you see it in action, you will recognize the signs. Sometimes the doe rejects the buck or is reluctant to breed. This happens most often with virgin does. The owner may hold the doe steady for the buck to mount. The buck may also be reluctant to mount. In the (fortunately rare) case of a slow buck, an alternate buck may have to be used. Sometimes teasing the doe with another buck on the other side of the fence induces a slow buck to mount her.

HEAT SIGNS IN THE DOE

- Allowing other does or wethers to mount her
- Calling out or crying frequently for no reason
- Exhibiting a drop in milk production
- Fighting
- Flagging (holding the tail high and frequently wagging)
- Acting more affectionate
- Losing interest in feed
- Mounting other does
- Exhibiting a mucus discharge from vagina
- Standing by the fence closest to the buck pen
- Exhibiting a swollen or pink vulva (rear end)
- Exhibiting tail hair that is wet, sticky, or clumped together

Taking your doe to another farm to be bred allows you to keep fewer bucks or no buck at all. Someone who raises only a few does should try to find a local breeder offering buck service. If you have registered animals, be sure to get a service memorandum so that you can register the offspring. Costs of outside breeding vary, but if you have only a few does, breeding or lease fees are usually less expensive than owning your own buck.

ARTIFICIAL INSEMINATION

To bring in new genetics, try artificial insemination (AI). You can keep frozen semen stored for many years in liquid nitrogen. AI is highly technical but readily learned by any goat owner with the interest. Unlike with cattle, there are few professional inseminators who come to the farm and breed your goats.

ESTRUS

Does typically reach puberty at about six months of age, although some become fertile sooner. Separate does from bucklings by three months of age. Most goats are seasonal breeders. The season is influenced primarily by light, with its onset triggered by decreasing hours of daylight. In the Midwest, the breeding season starts in September or October and runs through February.

Keeping the does and bucks penned separately and introducing them just before you want to breed will enhance breeding behavior. This introduction of a buck to a group of open does is called the "buck effect." Even off-season, does may respond to the presence of a virile male. If you want does to signal heat without the buck being able to automatically breed, a separate pen next to the does and downwind can cause the same effect.

Some breeds are easier to breed out of season than others. Myotonics, Boers, Pygmies, and Nubians are known to breed year-round. Individual animals may be exceptions, and some breeders select their goats for out-of-season breeding. During breeding season, does cycle into estrus every eighteen to twenty-one days. The first step in breeding is to have a doe in standing heat. This heat may last only a few hours or up to three days, depending on the animal and the time of year.

METHODS OF BREEDING

BREEDING METHOD	ADVANTAGES	DISADVANTAGES
Pen Breeding	Easy Not necessary to track heats Good for short-cycling does Good for hard-to-detect heats Useful for off-season breeding	May not know exact breeding date Difficult to perform prenatal care on doe without knowing breeding date May require additional pens, depending on number of goats being bred and number of bucks used
Raddle Harness	Similar to pen breeding Visual indicator of breeding	Expense of harness and markers Attaching harness may be tricky Buck may remove marker False marks with aggressive bucks or passive does
Hand Breeding at Home	Know exact breeding date May require fewer pens	Need good heat detection Requires close monitoring of does several times a day More time-consuming Handling the buck may be difficult
Hand Breeding off Farm	Similar to home breeding Wider range of bucks to choose from Eliminates or reduces the cost of owning bucks	Similar to home breeding May require multiple trips Travel distance and time May be difficult to locate breeder offering service Breeder may require certain health tests Possible to bring home diseases
Artificial Insemination	Wider range of bucks to choose from Faster genetic improvement possible Eliminates or reduces the cost of owning bucks	Initial equipment expensive Usually done by goat owner Good heat detection required Conception rates vary Conception rate dependent on technique

Sticky hair around the tail is one sign of estrus in the doe. *Jen Brown*

The smellier the buck and the better you rub his scent onto the rag, the better the lure of your "buck rag." *Jen Brown*

Once the doe ovulates, the egg has a viable life of about ten to twelve hours. Some does are clear about their desire to mate; others are shy or have quiet cycles. Goat keepers use several methods to detect heats in the herd. An otherwise healthy hermaphrodite goat is incapable of breeding and producing young. However, these animals are sometimes used as teaser goats to detect estrus. Because they often act bucklike, hermaphroditic goats react to does in heat just as an intact buck would—signaling to the owner when a doe is ready to go to a suitable mate. A vasectomized buck can serve the same purpose since his hormone levels remain high, unlike those of the castrated whether.

If there is no buck on the property, owners may use a "buck rag" to help detect heat. This method requires a rag, a jar with a tight lid, and a trip to visit a buck in rut. Rub the rag all over the buck's head and belly. Put the rag in the jar and cover tightly. Back home, open the jar under your doe's nose several times a day. When the doe gets excited or tries to get into the jar, it is time to visit the buck.

In herds that need out-of-season or timed kidding, does can be brought into heat using medication available from your veterinarian.

VISIT, LEASE, OR OWN?

Bucks expend a tremendous amount of energy breeding. Some bucks go off feed and lose weight during breeding season. Their single-mindedness can be dangerous. Most health problems with bucks occur during this time. Not keeping a buck reduces feed costs, the vet bill, and housing needs and prevents the occurrence of unwanted breedings should the buck escape his "escape-proof" enclosure. For these reasons, some owners elect to visit or lease the services of a buck rather than keeping their own.

The breeding season is hard on bucks. Fighting may lead to bloodied heads, and the energy expended mating depletes their reserves. *Jen Brown*

VISITING THE BUCK

Long before breeding season, you should talk to breeders in your area who offer buck service. Take into account pedigree, cost, distance, and herd health. Some breeders offer boarding for does, but more commonly the doe owner brings

There are many factors to consider when deciding whether you want to own, rent, or lease a buck for your breeding program. Cost, convenience, and housing and fencing needs are just a few. *Jen Brown, Poplar Hill Dairy Farm*

CHARACTERISTIC OF THE BUCK

TRAIT OR BEHAVIOR	CHARACTERISTICS	BREEDING FUNCTION	NEGATIVE ASPECT
Aggression	Rears up Butts Attacks fence separating him from does Challenges other animals Mounts other bucks or animals	Attracts does Establishes position in herd	Hazardous to caretakers and pen mates Requires strong fencing Need to separate horned and dehorned bucks to avoid injuries
Frequent erections	Exposes the end of the penis Inserts penis into mouth	Asserts dominance Attracts doe	Disconcerting for some owners Embarrassing for visitors
Increased verbalization	Moans and "talks" to does Blubbers and wags tongue	Attracts doe Encourages doe to stand	Sound can carry a great distance and disturb neighbors
Lip curling (Flehmen Reaction)	Lifts upper lip into a grimace Accompanies urine "tasting"	Exposes a sensory organ in the upper lip Detects estrus pheromones	Odd appearance
Scent	Gives off strong "goaty" odor Rubs scent on anything or anyone within reach	Attracts the doe Stimulates the doe to come into heat	Difficult to remove from skin and clothing May affect milk quality if buck runs with the does Offensive to some people and downwind neighbors
Urination	Urinates on front legs and face Sprays urine into mouth Sprays urine indiscriminately	Further attracts does by increasing pheromones	Urine scald on face and legs may leave buck's skin raw and irritated Can cause goat owner to change clothes (yet again) after being caught in the spray

the doe to the buck when she cycles. Because heat times vary among does, watch the doe and record how long her cycle lasts.

Biosecurity concerns have made it more difficult to find breeders offering buck service. Conditions that would prevent your doe from showing at a fair (abscesses, ringworm, lice, respiratory illness, diarrhea) should also prevent you from taking her to someone else's farm. Also, let the owner know if your doe has horns. Some breeders will not expose their buck to the risk of goring. The reverse is also true.

Have all registration papers and health certificates handy. Some breeders ask for these and check tattoos or ear tags before giving you a service memorandum. Expect to pay up front for a breeding service. Most breeders allow you to bring the doe back for another breeding at no charge during the same season if she does not become pregnant. Buck owners cannot be held responsible for does that abort, reabsorb the fetus, or do not settle unless a health inspection shows that the buck is infertile or is carrying a specific disease that caused the termination of the pregnancy. A breeding contract protects both owners from any misunderstandings.

The polite doe owner watches carefully for the first sign of heat and communicates its arrival as soon as possible to the buck owner. Actual breeding may take only minutes, but preliminary courtship and the paperwork following the act can be time-consuming. Be sure to factor "visiting time" between the goat owners into your schedule as well. Showing up at the farm without warning or insisting on coming at a bad time for the breeder may cost you the opportunity to breed there again.

LEASING A BUCK

Some breeders find it more convenient, especially if multiple does are involved, to lease the buck rather than take multiple trips to the buck farm. Buck owners may require an inspection of the leasing premises and additional health-status information. An agreement should be reached beforehand about what happens if the buck becomes ill, injured, or dies. Usually, the leaser feeds, houses, and cares for the buck. This may include covering veterinary costs if the animal becomes ill.

For pedigreed animals, goat associations have buck lease forms that can be filed with the association. The leasing farmer may then fill out registration papers and submit them without service memos. Service memorandums are required for any breeding that does not have a lease form on file at the office. Some breeders request that only does be registered out of the breedings. This arrangement should also be discussed before breeding or lease.

The older the buck, the more difficult it may be for him to relocate to a new farm and adjust to new feeding or housing arrangements. A buck that was content at home can go off feed, become ill during transport, or pine for his herd mates. I generally lease young bucks only to herds I know well or ones with only a few does.

OWNING A BUCK

If you decide to keep one or more "boys," it is important to know what to expect. The buck is very different from the doe. Even as kids, some bucks show breeding behavior. Intact male goats become increasingly bucklike as they age and hormones kick into effect. In certain breeds and in locations with strong seasonal effects, the buck's rutting behavior ebbs and flows with the season. In other breeds or in temperate zones, the buck may remain in rut most of the year.

The smell that most people associate with goats is the smell of the buck. This smell comes from oils secreted by scent glands located mainly at the base of the horns. This scent can be very strong and offensive to some people. Angora bucks have the scent gland not at the base of their horns but in their wool, so their smell may not be as strong.

In Greek mythology, satyrs are half-goat, half-man creatures and figures of fertility. This is no accident. The ancient Greeks knew their goats. Anyone observing bucks for any length of time sees that they are obsessed with breeding. During rut, bucks are odiferous, noisy, and aggressive. In other words, they stink, blubber, and generally act rude and obnoxious!

The chemical bromine, which has a strong odor and stings the eyes and irritates nasal passages, was named after the Greek word *bromos*, meaning "the stench of he-goats." This stench stays on anything it touches, including any parts of your body and clothing that come in contact with the buck. You will, at some point, need to remove this scent to be socially respectable. Cleansers that are supposed to work include Fast Orange Hand Cleaner, goat milk soap, Listerine, and toothpaste. Febreze in the wash is a must after working with the boys.

BYPASSING THE BUCK: ARTIFICIAL INSEMINATION

To render the above complexities irrelevant, consider AI as a means of breeding. Unfortunately, caprine semen processors are limited in number. Buck semen is usually collected at on-farm visits by the processor, who may visit a region only once a year. When an individual farm doesn't have enough bucks to make the stop affordable, clubs sometimes organize a "buck collection," which functions both as a social get-together for goat owners and an opportunity to semen shop from the processor's collection.

There are no professional goat inseminators in most areas. Goat owners learn AI through classes, books, and other goat owners. Classes are taught at shows, conventions, and other gatherings of breeders. Unlike with cattle, semen is introduced to the goat through a speculum and sight rather than by feel. It is an art as much as a science, with widely variable success rates between breeders.

PLANNING AHEAD

Plan ahead for the kids. As you read about raising kids, neutering, disbudding, and marketing, ask yourself the following questions:

- Will the kids be bottle-fed or raised by the dam?
- What vaccinations will I need?
- Will any buck kids be kept from this breeding?
- Will any bucklings be wethered, and how will that be done?
- Do the kids need to be disbudded, and how will that be done?
- How many kids will be kept in the herd, and how many will be sold?
- What type of markets are there for doe, buck, or wethered kids?

A buck in rut courts does by wagging his tongue (left). Another facial move is to curl the front lip, exposing scent glands—called the Flehman Reaction (right). *Jen Brown*

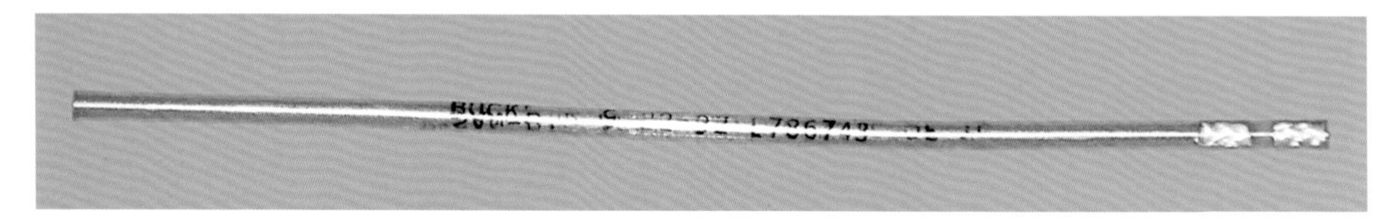

The most common storage for goat semen for artificial insemination is ½-milliliter straws. *Carol Amundson*

Buck semen freezes well and may be stored for years in liquid nitrogen tanks for artificial insemination use. *Carol Amundson*

IS MY DOE PREGNANT?

The first sign of a successful breeding is that the doe will fail to come back into heat eighteen to twenty-one days after the mating. Breeders say that the doe has "settled." I have known does who continued to seem to cycle after settling to the first breeding. For the owners of a few goats, it is helpful to know the doe is pregnant, if only to watch for those kids. In commercial goat operations, the sooner you know that a goat has settled, the better you can assess the breeding efficiency of your operation—are your does or your buck fertile, how is your breeding plan working, and so on. For any owner, the more you know about whether a doe is bred, the better you can plan drying off, increases in feed to support growth of the fetus; or vaccinations for passive transfer of immunity to the kids or prevention of abortion.

Goat owners have a variety of options besides merely waiting for the 150 days gestation that it takes to make baby goats. Some breeders claim to be able to tell pregnancy through observing puffiness in the vaginal area, or subtle changes in the doe's look. I have never found that I could tell a doe was pregnant by looking at her. On the other hand, a pregnant goat generally has a greater appetite and becomes calmer, quieter. The abdomen will increase in size, becoming rounder. For older goats with very large rumen, this can be tricky, as the rumen both increases and decreases in size with digestive gases as well as moves during the digestive process. In physically checking for pregnancy, theoretically at 6 weeks, a goat's abdomen diameter will increase by 1 inch or more. This increasing size can be distinct after 12 weeks or so.

More reliably, blood, milk, or urine testing has become more common and is available through your vet or by mail order. The earlier tests measured progesterone, a hormone that increases in pregnancy, but has been known to be variable and not always accurate. Pregnancy Associated Glycoproteins (PAGs) are specific hormones that are present in goats 28 days post-breeding. IDEXX has a milk test kit for PAG testing. The blood test BioPRYN measures the presence of Pregnancy-Specific Protein B (PSPB), a protein only produced by

Have a plan in place for what you will do with the kids once they arrive. *Jen Brown*

the placenta of a growing kid or lamb. Other tests, like P-Test, measure estrone sulfate (another pregnancy hormone). A number of labs perform this testing or offer kits, and that number is growing all the time. Because I have no direct experience with these tests in goats, I recommend you do research or rely on your vet if you plan on trying this.

Ultrasound is useful when a doe was pen bred or if there is some doubt about pregnancy. Forty-five to sixty days following breeding is the best time to check for pregnancy using this method. A relatively inexpensive machine sold to sheep and goat breeders is the Doppler ultrasound. This machine gives a way to measure fluids in the uterus to determine pregnancy.

Some vets own a real-time ultrasound machine, the same type used for human ultrasounds. It is exciting to see the ribs and beating hearts, count kids, and even estimate time of gestation. My vet and I got very good at measuring the distance between the eyes in kids. I would then use the measurement on a chart to estimate when the doe was bred and the kid was due. A blood test won't tell you this, and you cannot count the kids without real-time ultrasound.

CARE OF THE PREGNANT DOE

At least sixty days prior to the due date, examine the doe for fitness and remove a milking doe from the production line. The final days of pregnancy are known as the "dry period." Some goats require longer dry periods, but all does need at least two months off between lactations for the kids to grow and develop. If it is part of your herd practice and your veterinarian recommends using a "dry cow" udder infusion—a solution of antibiotics to help prevent infection or cure any undetected issues—do so on the last day of milking.

Start slowly increasing the amount of grain in the doe's ration or give her better hay as her condition warrants. Some dairy farmers bring the doe onto a milk stand twice a day to accustom her to the routine if she has never milked before, to examine her for problems, and to give specific rations.

In rare cases, a doe continues to milk in spite of the best efforts of the owner to dry her off. Some goats milk through their pregnancy and do not produce colostrum, which is normally the first milk produced by a new mother. Colostrum is thicker than regular milk, yellowish in appearance, and full of proteins and antibodies that protect the kid until its immune system is fully functioning. In its absence, you will need to take special steps (detailed on page 104) to ensure the kid's survival.

Thirty to forty-five days before the blessed event, give the doe annual vaccinations and shots. Ten to fourteen days before the known due date (or two weeks before the first possible date if the doe was pen bred), the doe should have a "pregnancy clip," also known as "crotching." This haircut gives the breeder an easier view of the back end of the doe to check for signs of parturition. Trimming around the tail and vagina keeps the doe cleaner at delivery. Trimming the belly and udder helps newborn kids find the teats and nurse successfully.

This is the time to move the doe to her separate kidding pen if you desire. Some goat keepers like to give the doe and offspring private quarters for delivery and bonding without interference from the herd. Other goat owners believe that the doe is less stressed when she stays in familiar surroundings. Certainly, if she is moved to a pen by herself, she should be within sight and sound of the herd to prevent loneliness.

Some farmers find a simple baby monitor useful. I used sound monitors for years,

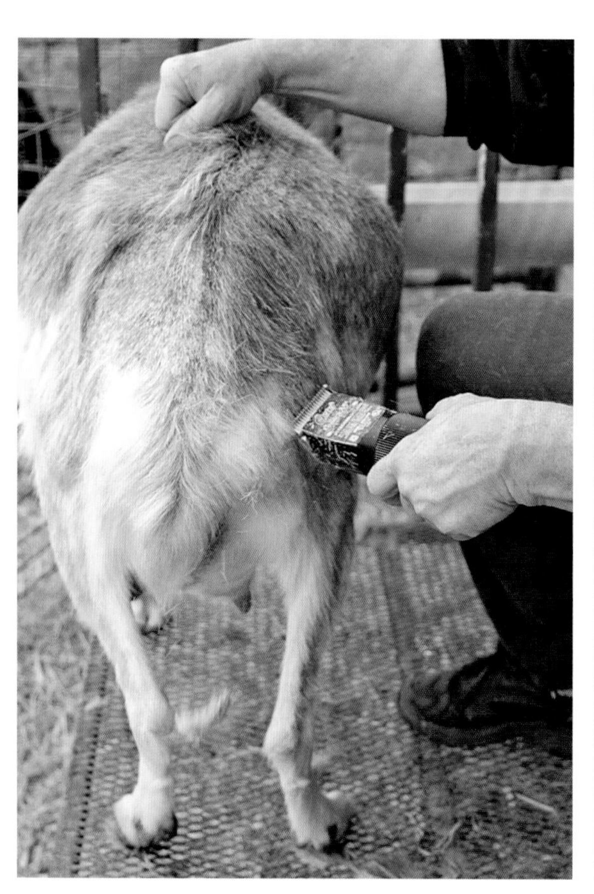

The pregnancy clip includes shaving from hips to pins (top) and shaving the udder (above). *Jen Brown*

PREGNANCY CLIP

SUPPLIES

Milk stand or stanchion, if available
Clipper lube
Small animal clippers
#10 blade
#30 blade

1. Lock the doe into a stanchion or milk stand, tie her closely to a fence, or have someone hold her.
2. With a #10 blade, shave down the rump from the hips to the tail. Be liberal with lube on the blades so that they won't get hot or gummy.
3. Moving up the tail, use a show clip so there are no hairs hanging on the sides closest to the vagina.
4. Shave the legs next to the udder, especially any shaggy hair on the back of the thighs.
5. Still using the #10 blade, clip any hair around the udder that would interfere with nursing or milking. This includes the udder arch, under the legs, and across the belly and fore-udder.
6. If you are concerned about ensuring a clean udder for the milk parlor, use a #30 blade to perform a close udder clip.
7. Give the doe a treat and release her.

although a barn full of groaning, snoring Nubians was enough to get me kicked out of the bedroom during kidding season. I learned to sleep through most sounds until a doe's pushing moans or the cry of newborns would rouse me out of a sound sleep. With the advent of reasonably priced video models, you can even see a view of the pen. Many farms now install computer security cameras in pens in order to observe goats through a computer or phone. Monitors are helpful for unusual situations or for detecting problems early.

SIGNS OF LABOR

There are a number of ways to tell that your doe is freshening, or approaching kidding. She may remove herself from the herd. Another doe may become more friendly and anxious to have her caregiver present. She may become vocal or testy with herd mates and others. One older doe of mine would toss any passing cat or chicken across her pen during the final hours before delivery!

You can watch for bagging up, or the filling and tightening of the udder with milk, but do not rely on this indicator. It is an inexact predictor, as some does develop a bag even when they haven't been bred. In virgin does, this is called a precocious udder. Other does don't let their milk down until after they deliver, although it is not common.

As the kid or kids move into position to be born, the doe's flanks hollow out, showing pronounced hip bones. The tail ligaments loosen so that the pelvis can widen as the kids pass into the birth canal.

THE DELIVERY KIT

The supplies you assemble to attend the birth may be as simple or elaborate as your comfort level dictates. Most of the time, a doe can deliver without help or interference. Other circumstances warrant special tools,

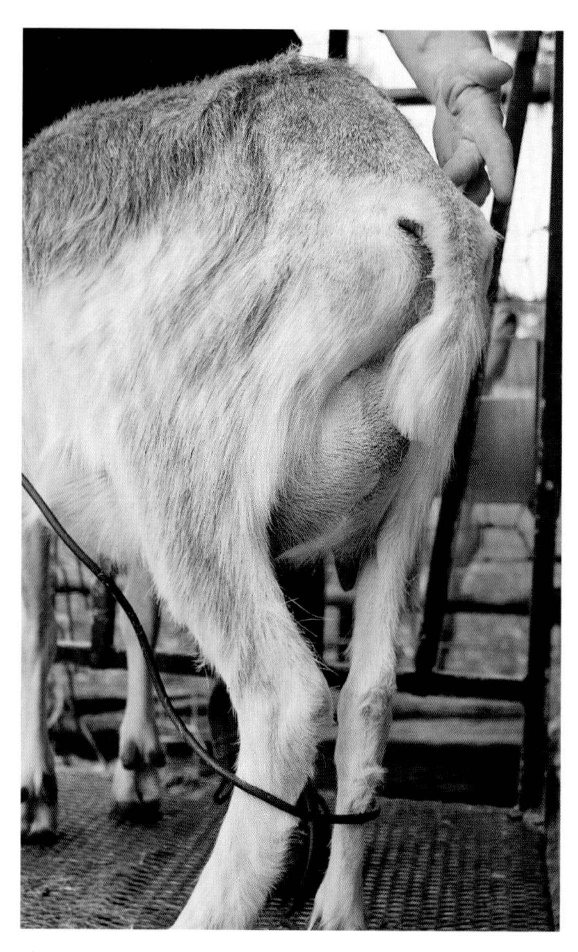

The pregnancy clip helps you detect hollowing out at the tailhead as kidding approaches. There are two ligaments beside the tail. These feel like tight bands most of the time. A sure sign that kidding is imminent is the hollowing out beside the tail. When you cannot feel the ligaments at all, expect kids. *Jen Brown*

medications, and supplements. The delivery kit list includes items that I have found useful over the years. Do not feel you need to have everything on this list. At times, my kit has been very small and didn't need to be larger. In some cases, however, it is better to be prepared than to wish you had an item when problems arise.

THE KIDS ARE COMING

Having babies is always exciting. I never get tired of welcoming new life into the world. That said, having a kidding kit ready and knowing what to expect goes a long way to make the experience positive. At the least, having supplies nearby prevents multiple trips to the house and panicked calls when you realize you don't have something you need. Since writing the first edition, I have beefed up the list of supplies and added other comments based on questions asked by new goat owners.

Your kidding kit is ready, and you have assembled the phone numbers of your goat mentor and your veterinarian. As the signs of kidding become more pronounced, the doe becomes more preoccupied (or starts kissing you and throwing chickens!). I had a doe that would attack anything close to her when she started early labor. No other goat, cats, or even chicken could get near her, or she would bite. Goats vary in their interest in having someone nearby. Most will look to their owner for help, or at least tolerate your presence. Because goats are a prey animal, wary does, particularly first fresheners, will get up, walk away, or otherwise try to hide their labor.

Probably the first sign you will see of the impending birth is a string of mucus hanging from the vulva. This can be a dense string of clear mucus that hangs all the way to the ground. If the mucus string is dark yellow or brown, be prepared to possibly help the doe. The color is due to meconium, or dark baby poop, and signifies stress on the kid. A sudden gush of fluid signifies that the bag of amniotic fluid the kid has been growing in has burst.

Natural labor can last as long as twelve hours in first-time mothers or does with multiple kids. Normally, the serious labor immediately preceding delivery should last about an hour—no more than two. After that, carefully check to see if the doe needs help. Gloves and lubricating gel allow you to stick a few fingers into the vagina to feel for the cervical opening, the bubble, or parts of the kid.

It is important not to wait hours or overnight once actual pushing labor starts. I have read too many sad stories online from new goat owners whose doe had malpositioned kids. The doe can exhaust herself or kids can die if not tended to immediately. It is better to be safe than sorry—check the cervix and kid position.

At this point, get your kidding supplies close at hand. Kneel on a tarp, newspaper, or empty feedbag in the pen for cleaner delivery. If you are delivering kids at night in a poorly lit barn

DELIVERY KIT

BASICS

Bucket of warm soapy water for cleaning off hands and doe
Colostrum—thawed frozen or fresh colostrum from another doe
Iodine 7% or Betadine or Nolvasan
J-lube, K-Y Jelly, or other lubricant
Nipples and bottles
Laundry basket or tub (to hold supplies and double as a kid holder)
Paper towels, old bath towels, or newspapers
Surgical gloves—long, disposable
Scissors
Warm water for doe

FOR WEAK OR CHILLED KIDS

Blow dryer
Digital thermometer
Feeding tube and syringe
Heating pad
Bo-Se or Selenium Oral Paste
Strong black coffee (also for the delivery team)

NICE TO HAVE

Baby monitor
C&D antitoxin
Calcium Drench
Calendar or notebook for records
Dental floss or navel clamps
Electrolytes—ReSorb, Pedialyte, or Gatorade (*not* diet)
Fortified Vitamin-B Complex
Flashlight or light source
Goat serum concentrate or colostrum substitute
Kid coats
Kid puller or leg snare
Kid tags or ID bands
Nasal bulb
Nutri-Drench, molasses, or energy supplement
Penicillin or other broad-spectrum antibiotic
Preparation H (relieves vulvar swelling)
Probiotic or yogurt
Propylene glycol
Puppy training pads
Syringes—10cc and 20cc with 20- or 21-gauge needles
Tarp or sheet of plastic
Uterine bolus

PRESCRIPTION MEDICATIONS FROM VETERINARIAN

Bo-Se/Selenium (veterinary injectable)
Dexamethasone (for labor induction or to help preemies' lung development)
Dopram (under-tongue lung medicine)
Lutalyse or estrumate (for labor induction)
Oxytocin (to stimulate contractions)

or shed or on pasture, set up lights or have a flashlight to help you see. As you catch the kids, focus first on clearing the kid's nostrils for easy breathing. Rub the kid's face briskly with newspaper or towels. A nasal bulb can be used to suction the nose. Once the doe is in serious labor, the kids come fairly rapidly. First-time mothers may take longer. Singles are usually born faster than twins. Triplets or more can be a problem because all those long legs get tangled up. In most cases, patience rewards you, and the kids reposition naturally as the doe paws the bedding, stretches, and gets up and down on her feet.

Eventually, you will see a bubble appear. This amniotic sac, called the "water bag," appears and recedes as birth fluids lubricate the passage and the birth canal widens. Toward the end, the bubble should burst and release fluids. As mentioned previously, you may see a gush of fluids with no sac. This is a pretty good sign that the kid cannot get through the vaginal opening in its present position and needs help. With luck, the kid's head has now crowned past the tailbones and its nose is clear so that it won't suffocate or aspirate fluid. Once the umbilical cord separates from the placenta, the kid no longer receives oxygen from the doe. The best position for the first kid to present is therefore with its nose tucked over the top of its toes.

Another normal presentation is both hind feet first, followed by the hips and tail. A second kid is often delivered this way due to the twins being nested one on top of the other, head to tail. A second common position for twins is for both to present front feet first. Sometimes kids in this position become tangled and need assistance.

Do not be alarmed if the doe cries out in distress. Most does make some sounds during labor, although some are silent. The doe may also "talk" to you in soft bleats that you will later come to recognize as goat baby talk.

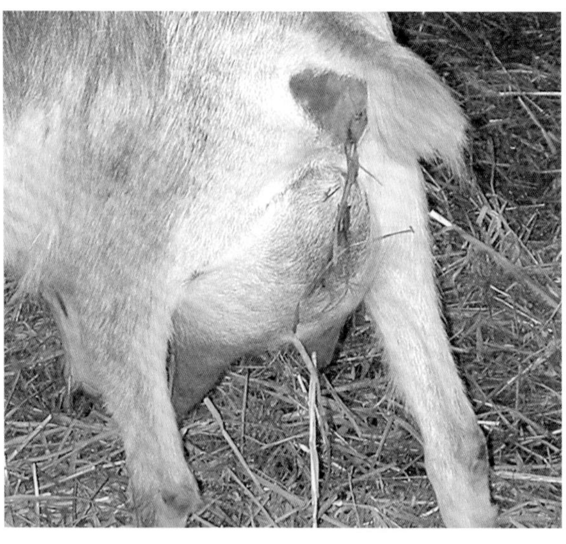

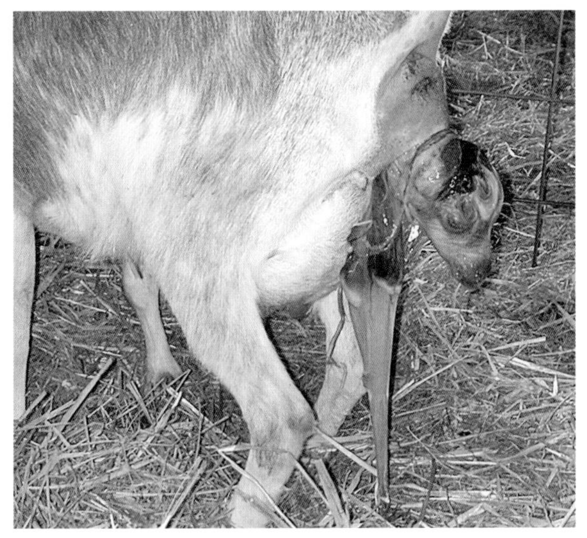

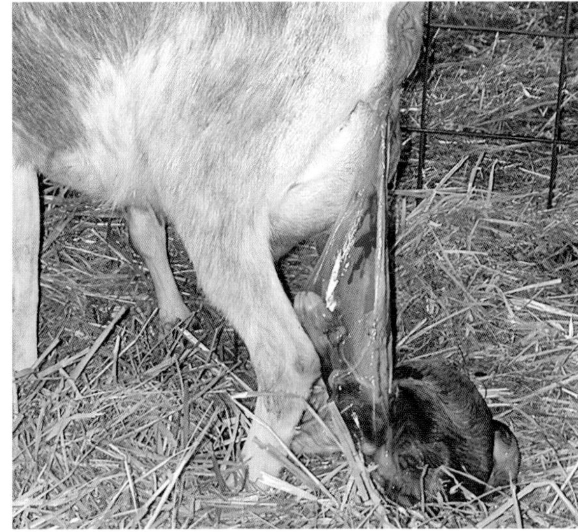

At some point in the birthing process, you will see a bubble appear. This amniotic sac will appear and disappear as birth fluids lubricate the passage and the birth canal widens. Toward the end, the bubble should burst and release fluids. With luck, the kid's head has now crowned past the tailbones and its nose is clear so it doesn't suffocate or aspirate fluid. Once the umbilical cord separates from the placenta, the kid has no more oxygen from mom and needs to be out in the air as soon as possible. *Carol Amundson*

For a number of years, I used a nonvideo baby monitor. Hearing a doe in labor was usually my sign to check the barn. However, sometimes, all that I heard was the kid crying.

After all the kids are born, the doe delivers the placenta, also called the afterbirth. Left to her own devices, she usually eats this nutrient-rich membrane. If the delivery results in stillborn kids, however, the placenta can be invaluable for diagnosing the cause. Use gloves to collect the afterbirth along with any fetal material. Place a questionable placenta in a plastic bag and refrigerate it. Contact your vet or the state veterinary lab for help diagnosing the cause of stillborn kids. The placenta is not always delivered right away; don't be concerned if it takes as long as 12 hours after kidding. Sometimes, it even delivers partway and hangs from the doe. It is not necessary to pull or break it off. Actually, the weight of the placenta naturally pulls the entire organ free.

If you choose to raise your goats on a strict CAE prevention program, the birth is your time to step in. Since CAE is transmitted in bodily fluids, clean and dry the kid off with towels or newspaper as briskly and thoroughly as possible. This cleaning also mimics the mother's licking and cleaning of the kid. The stimulation arouses the kid, helping it to breathe normally. If you think it has swallowed fluids or if you find the kid still in the birthing sac, quickly clear the nostrils of membranes and mucus. It may be necessary to hang the kid upside-down by its hind legs and gently swing it to expel swallowed birthing fluids from its lungs. Use your bulb to clear nostrils if you have one.

Once the kid is breathing well, milk out the colostrum from the doe and save it for the kid or freeze it as an emergency supply. Before giving the colostrum to the kid, be sure to follow the directions for heat-treating found later in this chapter. Colostrum substitutes do not always contain antibodies, so I don't tend to recommend them. Check the product label or insert if you are unsure.

Kids can be put into a laundry or storage tub with newspaper or puppy pads in the bottom. This works for both dam- and bottle-raised kids and gives them a safe place to dry off without wandering. Some kids are up on their feet very quickly and will skitter away and hide. The tub allows you to separate the kid from mom and her fluids or you can set it next to mom, so she can lick the kids there. Of course, very active dam-raised kids should be nudged into place to nurse rather than spending tub time.

If you choose to allow the dam to raise her kid, let her lick the newborn clean and start to nurse. Constant nuzzling, sniffing, and contact are critical to forming a bond. Caprine bonding is mainly based on scent. From the moment of bonding throughout life, the doe and kid will recognize each other.

Sometimes, though, this bonding is disrupted. Kids that must be separated from the mother at birth for medical reasons but are returned within two hours will usually be accepted by the dam. If it is very cold in the barn, I like to be sure the kid is dried completely using my blow dryer so the ears and feet won't freeze. I still make it a point to get dam-raised kids back to mom ASAP. After three hours of separation, the bond can be difficult to forge. Sometimes rejected kids are accepted by a doe if a strong-smelling substance, such as menthol, is put on the kid and the doe's nose. Resistant mothers may form a bond during forced exposure in a separate pen; however, this can take up to ten days, if it works at all.

When rejection cannot be overcome, you may find help elsewhere in your herd. Some young, subordinate does allow other kids to nurse them. Other does have very strong maternal instincts. These goats willingly raise another goat's kids that are given to them or may even try to steal other does' kids. One warning about extremely maternal does: I lost kids when a doe tried to steal another goat's kids at birth. The kids were trampled during the event. Some people have separate kidding pens for their pregnant goats during and after delivery. The privacy gives a good chance for bonding without the rest of the herd getting involved.

ABNORMAL BIRTHING SITUATIONS

Sometimes things just don't go as planned. Intervention can be required in the case of a very large kid being birthed by a small doe, tangled kids, or a breech delivery, which occurs

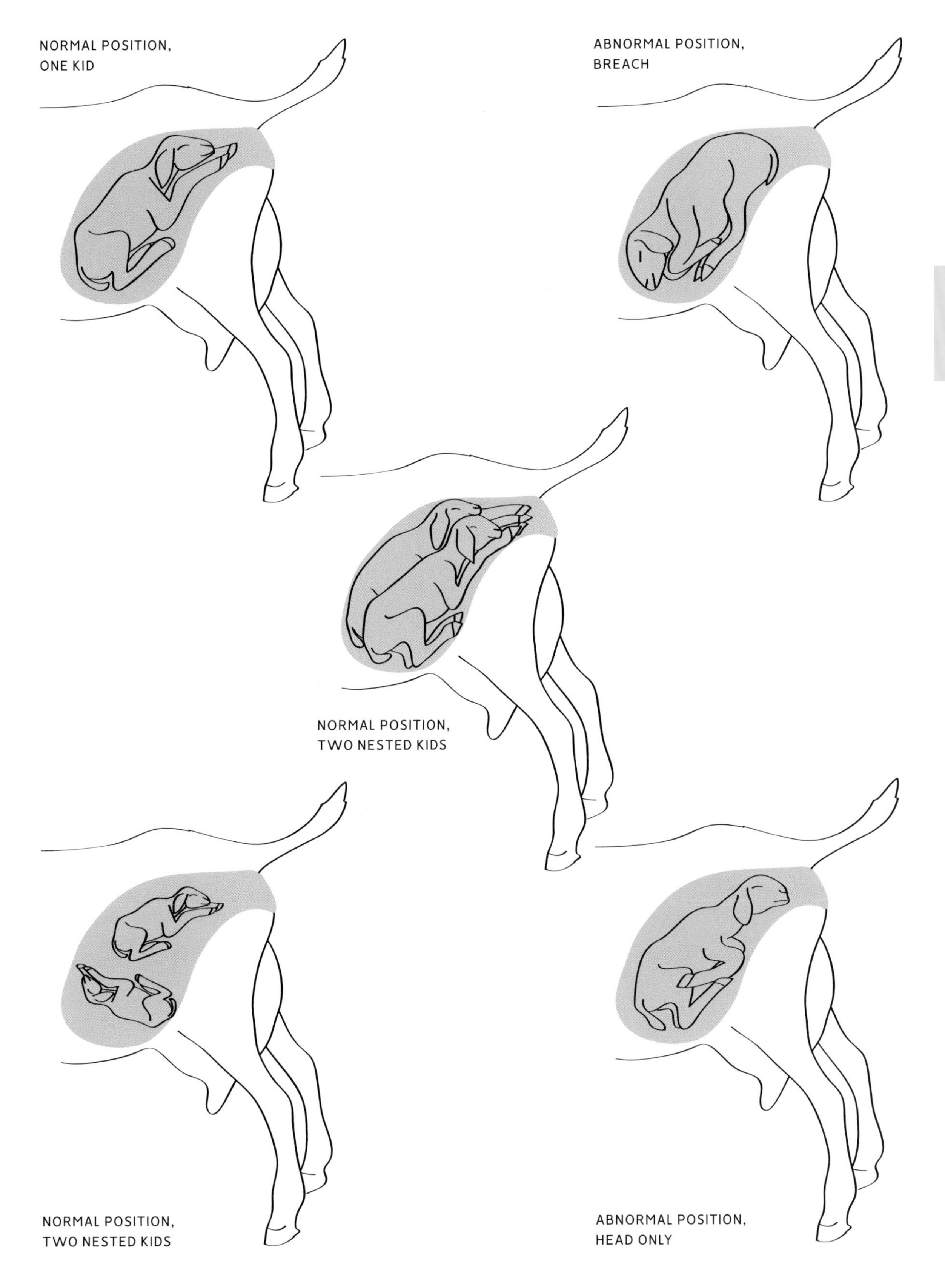
NORMAL POSITION,
ONE KID
ABNORMAL POSITION,
BREACH
NORMAL POSITION,
TWO NESTED KIDS
NORMAL POSITION,
TWO NESTED KIDS
ABNORMAL POSITION,
HEAD ONLY

The doe licks and sniffs the kid to develop the maternal bond. *Jen Brown*

when the kid's backbone or tail blocks the birth canal. This type of event is when you need your vet or mentor, if possible. Delivering stuck kids is not for the faint of heart. On the other hand, with good references and courage, anyone can come to the aid of a doe.

I had firsthand experience of this when my then ten-year-old daughter delivered tangled kids on the kitchen floor. Recovering from back surgery and banned by doctor's orders from the barn, I was not able to assist birthing. A doe was in trouble, so my farm partner Dave brought the animal to the house for my advice. Viveka had seen me pull kids before. With her smaller hands and my directions, she helped successfully deliver two kids, although unfortunately, one was dead. The second kid would not have survived without removing the deceased twin in a timely manner. Before placing your hand inside the animal, wash your hands with Nolvasan or another disinfectant to prevent infection. Your fingernails should also be short to avoid injuring the uterine wall. Wearing gloves helps protect you from disease and keeps the inside of the doe as clean as possible. Use a good lubricant. J-Lube, a powder that becomes extremely slippery after mixing with water, is great. In very difficult cases of stuck kids, use a kid-feeding syringe to put lubricant directly into the birth canal.

People with larger hands may need to use a kid-pulling loop or a leg snare. I have no luck with these implements and prefer to feel the kid and uterus by hand. Working inside the animal is crowded. It is hard to identify the kid part or parts you are touching. You may be able to picture things better by closing your eyes as you feel inside the doe.

Despite any fear you have that you will hurt the doe, remember that the kid is at least as large as your hand, even in miniature goats. Work slowly and cautiously. You may find that your hand is squeezed as the doe continues to have contractions. Do not be surprised if you have a bruised hand later.

Some situations are simply beyond the goat keeper's capabilities. When necessary, call the vet for professional help and perhaps a cesarean section. Birth defects, such as two-heads or kids that die before delivery, may require that the fetus be cut from the doe. Luckily, such defects are extremely rare. After assisting in the birth, be sure to clean your hands thoroughly, since some of the diseases goats carry are transmissible to humans. There also have been cases of people, usually veterinarians, becoming sensitive to birthing fluids and developing an allergic response.

CARE OF THE DAM

Your doe has been through a difficult time. She will appreciate a warm bucket of water after delivery. Some owners add molasses to the water for energy and iron. Others give goat Nutri-Drench, electrolytes, or other boosters. Be careful about using Gatorade. There are so many varieties on the shelf that it is easy to grab a bottle that has artificial sweeteners or sorbitol. Artificial sweeteners are not safe for your doe and may do more harm than good.

If you had to assist in the birth by reaching into the doe, she'll need protection from any contaminants that may have been introduced to her womb. Hopefully, you have used gloves and tried to be as clean as possible. A uterine bolus and/or a broad-spectrum antibiotic can stave off infection. Insert half a cow bolus deep into the doe. Some breeders give injectable penicillin to the mother for five days following delivery. If she is torn or badly swollen, try Preparation H liberally slathered over the vaginal region to make her more comfortable.

The day after kidding is a good time to administer a wormer. During the last month of pregnancy and at kidding, the pregnant doe is under great stress. This stress causes her immunity to decrease and the parasite load to increase—releasing many eggs. Occurring over about four weeks, this egg release is the primary cause of new parasitic infections in kids.

CARE OF THE NEWBORN

If the kids are staying with their mom, you don't have too much to worry about. Typically, a doe cleans off her newborn quite well on her own. In colder weather or when she is having multiples, however, it is good to help her out with newspaper or old bath towels. If the weather is below freezing, use a blow dryer to prevent exposed body parts from freezing. The most vulnerable areas for freezing are the ears, tail, and feet. In my neck of the woods, there are Nubian or Swiss-type goats that we refer to as "Minnesota LaManchas" because their ears have frozen off! Angora goat kids are especially sensitive to chilling, so make sure they are warm and dry.

Kids that will be hand raised on a CAE prevention program need to be brought into another area of the barn or even the house. Laundry baskets or tubs make good nurseries for the first day or so. *Carol Amundson*

CARE FOR THE UMBILICAL CORD

The umbilical cord tears as the kid is expelled from the uterus. It can be any length. This tearing will normally stop the blood flow, making tying or clamping unnecessary. Use a surgical scissors to cut the cord to 2 inches if it is too long.

Once you cut it—or if you choose to leave it as it tore—you should dip the cord in 7 percent iodine solution, Betadine, or Nolvasan disinfectant. This procedure helps prevent navel ill infection and dries the remaining tissue. The iodine also prevents bacteria from growing in the open wound while the cord dries out. Pouring the iodine into a small disposable cup will make dipping easier.

Rarely, the cord is very thick and seems as though it won't stop bleeding. In this case, tie it off with dental floss or thread. You can also use the umbilical clamps sold by livestock supply companies. Dip the clamp or floss in iodine and then dip the tied cord again.

CHECK OUT THE KIDS

Examine the newborn kid carefully. Is it a buck or a doe? Don't look too quickly; even experienced goat keepers have been embarrassed to discover later that their newborn doe was a buck or vice versa!

Look at the jaw and palate. Very rarely a kid is born with an open palate, which is a palpable hole in the roof of its mouth. These kids can't suck properly and usually die or must be put down. A severe overbite, also called a parrot mouth, is undesirable as well.

Examine the teats on both does and bucks. Ideally there should be only two. Clusters or other abnormalities may cause difficulty for the does later when milking or nursing kids. Bucks with extra or split teats should not be kept for breeding. Missing testicles or undescended testes, known as cryptorchidism, are also criteria for culling a buck. Other birth defects are always possible, so look the kids over completely.

IDENTIFICATION

If many kids are born on your farm, be sure to mark them all at birth so you know which dam had them. Marking is even important for dam-raised kids, since does have been known to steal kids from one another. In pasture or group housing, kids grow quickly. Identifying now saves confusion later.

Initial ID of kids can be as simple as a colored mark on their head or body with a matching color on the mom if they are range animals. Temporary paper or Velcro ID bands

After a few days, kids become active enough to jump or climb from their tub. At Poplar Hill Dairy, rambunctious kids are restrained until they can be moved to other housing. *Jen Brown*

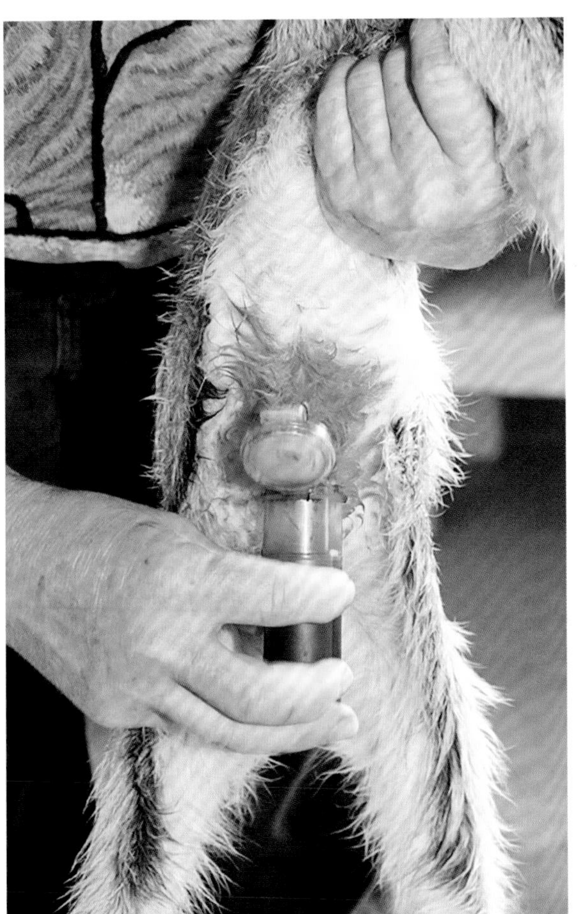

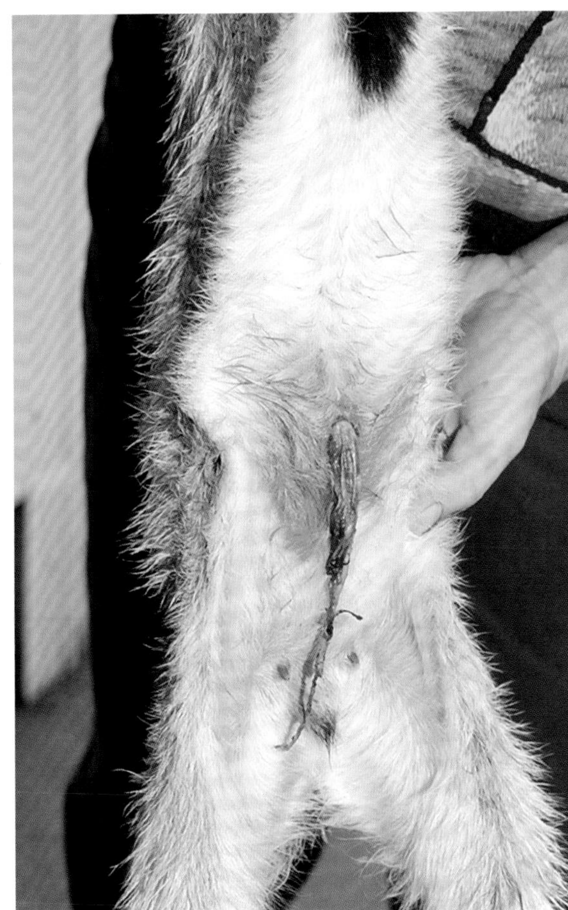

Dipping the navel with iodine or Nolvasan keeps the newborn healthy. *Jen Brown*

may be purchased. Breeders put the birth date, dam, sire, and other information on the band before attaching it around the kid's neck. Some goat keepers just use collars and tags right away. Tattoos stay clearer and last longer if you apply them when the kid is a few months old, so don't tattoo newborns.

INJECTIONS AND IMMUNE SYSTEM SUPPORT

Kids from unvaccinated mothers will require the C&D antitoxin to prevent intestinal problems associated with enterotoxemia. Kids should also receive a tetanus antitoxin. The antitoxins provide protection for about three weeks. In selenium-deficient areas, kids also need a Bo-Se injection or small amount of oral selenium gel.

Don't give the C&D and tetanus vaccines at this time, however, as the kid's immune system isn't developed enough to make use of the shot. Kids should be vaccinated for C&D and tetanus at one month and again at two months.

DAM-REARING VERSUS BOTTLE-RAISING

Kids can be raised by the dam or bottle-fed. Each method has breeders strongly on one side of the fence or the other. When I started raising dairy goats, I would never have considered dam-rearing except in the case of market kids. Now, after adding meat and pet goats—and finding schedule and health problems affecting the time I'm able to spend with the goats—my kid raising is strongly geared toward dam-rearing.

Dam-rearing is the least labor-intensive method of kid rearing. Most often used by the breeders of meat, fiber, or pet goats, this method requires very little attention from the owner. Be aware that some customers have a strong preference for people-oriented,

Dam-rearing (left) and bottle-raising (right) both have their merits. *Jen Brown*

hand-raised kids. Especially in the dairy goat market, some buyers are adamant about buying kids raised on disease prevention protocols. Opponents of dam-raising have strong views on the matter. Dam-raised kids are said to be less people-oriented and less friendly than their bottle-fed counterparts. More importantly, from the standpoint of animal health, dams may transfer one or more of several diseases from their milk to their kids. CAE and Johne's disease are two diseases that can pass in milk and remain latent, yet possibly infective, in the offspring. The kids run the risk of catching parasites, bacteria, and coccidia while running with the herd or even alone with their mom. Dairies may require the doe's milk for the bulk tank and prefer to find alternate sources of milk for bottle-raised kids.

On the other side, fans of dam-raising speak highly of the process, citing the fact that the kids are raised as "nature intended." Advantages include faster growth rates, earlier consumption of hay and grain, and easier integration into the herd. When separated along with their dam into small pens for the first few weeks, the kids become socialized to humans as well, so they still make fine pets.

DAM-REARING

Most goats are excellent mothers, and instinct is a wonderful thing. The best way for the dam and kid to bond is if they are allowed to be by themselves in a separate pen or a corner of the barn. It is initially important to check kids frequently to ensure that they are getting enough to eat. Once a kid stands, helping it find the nipple and even squeezing out the plug at the end of the teat gives it a good start.

If the doe is slow to let down her milk or if a kid is weak or slow to eat, be aware that it is critical that newborn goats get colostrum as soon as possible after birth, ideally within the first hour. Newborns have less than twenty-four hours to receive colostrum or their systems cannot absorb its antibodies. Optimum time frame is within six hours. The first-day colostrum from a doe has the highest level of antibodies.

If needed, you can help things along by giving fresh or frozen colostrum. Supplemental bottles do not affect the bonding between the doe and her kids. A general rule of thumb is to feed 10 percent of the kid's body weight in colostrum within eighteen hours. An 8-pound kid thus should get at least 13 ounces of colostrum. Because heat-treating fresh colostrum takes a little over an hour, I try to keep heat-treated colostrum in plastic pop bottles in the freezer for newborns. When I know a doe is in labor, I pull a bottle from the freezer and warm it in a bowl of hot water while waiting for kids.

Does with triplets may need help providing enough milk for their litter. When penning a group of does with kids together, do not put the smaller, multiple-birth kids and their moms with larger, single kids. Larger kids have been known to steal milk from weaker ones.

BOTTLE-REARING

In addition to their immediate need for colostrum, kids should get 10 to 15 percent of their body weight in milk each day. This is ideally divided into three to four feedings timed equally apart. Some breeders find that twice-daily feedings are enough. The kid's tummy is a good indicator. When it is obviously full and slightly distended, he has had enough—no matter how much he thinks he should keep drinking. Kids should be weaned at eight to twelve weeks, once they are eating hay and chewing a cud. Make fine-stemmed grass hay and fresh water available all the time from about one week of age.

PAN FEEDING

Some breeders teach kids to drink out of a pan. This works well, with a few caveats. Kids may stick their feet into flat pans and dirty the milk. In cold weather, pan feeding may be a problem for Nubians and other longer-eared goats that get their ears wet as they dangle in the milk.

This dam is still nursing her kid at five months of age. *Carol Amundson, Cutter Farms*

CAE PREVENTION AND PASTEURIZING

After caprine arthritis encephalitis was first identified as a serious problem in dairy herds in the early 1980s, it was also discovered that heat-treating and pasteurization would kill the virus. Since CAE has no cure, the practice of CAE prevention was developed. This model has drastically reduced the number and severity of CAE cases in the United States. In fact, feeding kids heat-treated colostrum and pasteurized milk prevents more diseases than just CAE, since it kills any disease organism present. (More details about CAE can be found in Chapter 4.)

HEAT-TREATING COLOSTRUM

The antibodies that colostrum contains are made of proteins that are sensitive to heat. If you attempt to pasteurize colostrum the same way you pasteurize milk, the proteins break down and coagulate. The resulting custard has none of the active protections that colostrum provides and cannot be fed through a bottle. Heating colostrum to a lower temperature and holding that temperature for a longer time kills CAE and other organisms without destroying the liquid's protective properties.

Colostrum should be heated between 133 and 138 degrees Fahrenheit for one hour. The best way to do this is using a double-boiler or water bath. USE CARE! If the colostrum is heated above 140 degrees Fahrenheit, the antibodies will be destroyed. Some breeders use canning jars in the water bath. Measure the temperature in the jar, not the temperature of the water, since it takes longer for the colostrum to heat.

A thermos can be used to hold the colostrum at the right temperature. Once the colostrum reaches 135 degrees, transfer it to a prewarmed thermos and set a timer for one hour. Rinsing the thermos with hot water before the transfer prevents the colostrum from cooling too much. At the end of the hour, check the temperature of the colostrum; it should not be less than 133 degrees Fahrenheit.

Heat-treated colostrum may then be poured into 20-ounce soda bottles, which work with either a Lambar nipple or Pritchard Teat. The bottles may be stored in a freezer, ready to be thawed and used for future kids. Defrost frozen colostrum slowly in a warm-water bath. Be careful not to overheat and ruin the colostrum. Don't thaw colostrum in the microwave, as it will heat unevenly, destroying valuable antibodies. Nothing is worse than turning that lovely colostrum into custard just when you have new kids to feed.

Multiple kids can eat from a Lambar, which is a bucket with nipples and tubes going into the milk (left). As with bottle-feeding, kids need to be trained (right). *Carol Amundson (left), Jen Brown (right)*

The dam bonds with the kid through scent. *Barb O'Meehan*

A polyurethane pig nursery works great for bottle-raised kids for the first few weeks, as it is safe and easily cleaned. As the kids get larger, they start popping up to see what is going on. *Jen Brown, Poplar Hill Dairy Farm*

Feed only real colostrum for the first feeding so that the kid gets the best possible protection. Colostrum substitutes are on the market, but none provide the antibodies that real colostrum does. Some breeders feed their kids cow colostrum, but be aware that diseases may also be present in cow colostrum.

PASTEURIZING MILK

Milk is pasteurized at a higher temperature for a shorter period of time than colostrum. Heat milk to 165 degrees Fahrenheit and hold for five minutes. This can be accomplished in a double-boiler or other water bath on the stove.

A raised mesh floor keeps the kids clean by letting wastes drain away or settle in the bottom. *Carol Amundson*

The SafGard pasteurizer and Weck canner are common commercial pasteurizers for home use. *Carol Amundson*

Commercial pasteurizers are available from dairy- and goat-supply stores. The most commonly used commercial pasteurizer is SafGard, which comes in a standard version and one that has a heat-treating setting for colostrum. The SafGard pasteurizer will heat up to 2 gallons of milk. Connected to a water source, this pasteurizer also cools milk rapidly by exchanging hot water for cold once the heating has been completed.

Another good-quality pasteurizer is the Weck Electric Canner. Carried by Khimaira Farms and Caprine Supply Company, the canner holds up to 5 gallons of milk, may be used as a water bath, and even holds a stainless-steel milker bucket. Because it is sturdy, simple, and well made, many breeders had been using this canner. Now, in 2018, I see that the Weck canner has been difficult to find in stock. A similar electric canner, available on Amazon, is the Ball Electric Canner. If you want to pasteurize using this type of equipment, you want to find a water bath that allows you to adjust the temperature and ideally put a container of milk inside to avoid scorching.

CAE PREVENTION KIDDING PROTOCOL

Tape teats of pregnant does one week before due date to prevent nursing if you miss the delivery. Place CAE-positive does in a separate kidding pen so they will not infect the other does via birthing fluids.

Remove kids from the doe immediately after birth, allowing minimum contact between the dam and kid.

Feed colostrum from a safe source or heat-treated colostrum within the first couple of hours after birth.

Feed pasteurized milk, CAE-free milk, or milk replacer until weaning.

Leaving kids in pasture lets them play in the sun. *Barb O'Meehan*

4 HEALTH AND WELLNESS

I DON'T KNOW HOW OLD I AM BECAUSE THE GOAT ATE THE BIBLE THAT HAD MY BIRTH CERTIFICATE IN IT. THE GOAT LIVED TO BE TWENTY-SEVEN.

—SATCHEL PAIGE

Healthy goats can live into their twentieth year, but the quote from Satchel Paige suggests either he exaggerated or his family should have consulted the folks at Guinness Book of World Records for their pet. According to Guinness, the world's oldest goat ever was McGinty, a pygmy goat who lived twenty-two years, five months. He died in November 2003 and was owned by Doris C. Long of Hayling Island, Hampshire, United Kingdom. I have personally known goats who lived close to twenty years, but this is rare.

BIOSECURITY

The word *biosecurity* calls to mind images of large, antiseptic factory farms. However, it applies to all operations, large or small, and can be the key to keeping healthy livestock. Any measure that decreases the chance of disease-causing organisms entering the goats' environment increases your level of biosecurity.

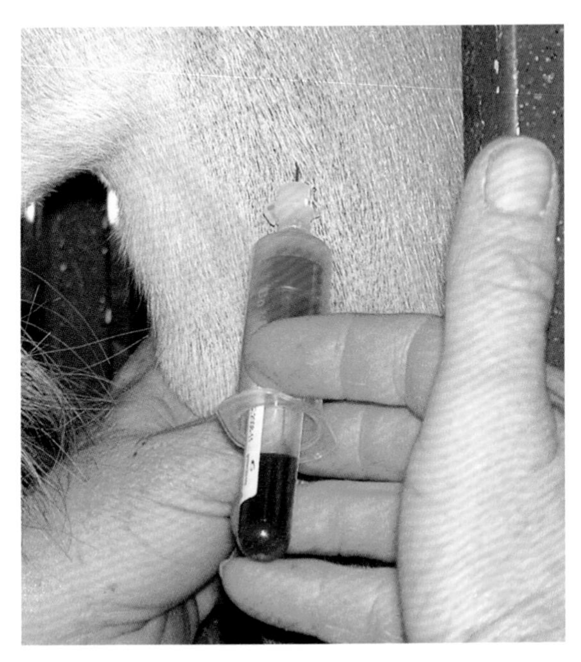

When performing caprine venipuncture, you need to secure the vein and then fill the tube. *Carol Amundson*

Each goat keeper needs to determine how far to take the following biosecurity measures:

- Clean and disinfect pens frequently.
- Limit or restrict bringing outside animals into the herd.
- Isolate new animals, sick goats, and those with chronic contagious conditions.
- Limit or eliminate any travel for goats, including shows.
- Change your clothes or put on coveralls when moving from regular pens to isolation pens.
- Wash your hands frequently between animals and before going from pen to pen.
- Change and disinfect your clothes and foot coverings when you come home from shows or other farms.
- Restrict visitor access to livestock housing and limit their handling of your goats.
- Request that visitors do not wear the same clothes from their barn to yours. (And give their farm the same courtesy.) Have a change of outer clothes or boots for them.
- Use gloves for invasive procedures or those involving body fluids.
- Disinfect or change equipment between goats when performing procedures that involve needles, tattoo digits, hoof trimmers, or other tools that come into contact with body fluids.
- Properly dispose of potentially biohazardous materials.

BLOOD TESTING

Goats may be tested for a variety of diseases and conditions. Buyers or owners may request testing for specific conditions like CAE to facilitate purchase or culling decisions. Your veterinarian can perform the blood collection. However, many breeders who use regular testing as part of their herd management find it less expensive and less troublesome to collect their own blood samples. I give instructions for collecting samples using a vacuum tube and holder system. If desired, the sample can be collected in a syringe and transferred to the vacuum tube.

Since I first wrote this book, caprine blood testing has become much more common. Whole-herd or individual animal testing is less expensive, making it more accessible for herd owners. Multiple labs now offer reliable tests for the big three—CAE, (caseous lymphadenitis (CL), and Johne's—plus pregnancy tests and genetic evaluations. Breeders use these tests to diagnose, make culling decisions, and promote their sale animals. The demand for "disease-free" or tested goats continues to rise.

CHOOSING A LAB

You can ask your veterinarian for suggestions about which laboratory to use for testing. Some labs require a veterinarian to be working with the herd. Other labs accept test requests directly from the farmer. Washington Animal Disease and Diagnostic Lab (WADDL) is one of the most respected laboratories performing goat testing. This lab is the lab of choice for many veterinarians and it also allows herd owners to send testing directly. Biotracking, Pan-American Vet Labs, and Sage Ag Labs are some other common labs that work directly with the consumer. The laboratory should be one that commonly handles goat testing and undergoes regular quality-control testing to demonstrate competency.

Different laboratories use different test methods. It is important to understand those differences. Read materials supplied by a testing lab to learn what type of test is used, what constitutes a good sample, and how to interpret the results. Every test method and each lab will have individual strengths as well as weaknesses. There is no perfect laboratory test. Tests can be misread through problems with the sample, problems with the kit, and the inherent percentage of inaccurate results present in any method. Even a mislabeling or typographical error is rare, but possible. I prefer to test but verify. It would be awful to cull a loved animal due to a single result.

The first CAE testing was performed using Agar Gel Immunodiffusion (AGID). AGID is very accurate, so there are few false positives, but it isn't very sensitive. False negatives are possible with AGID testing in early disease stages when the level of antibodies is low. Enzyme Linked Immunoassay (ELISA) tests detect lower levels of antibody and have fewer false negatives. Both are indirect tests, measuring the antibodies to the infective agent.

Polymerase Chain Reaction (PCR) testing is DNA-based. PCR tests for the genetic material of the actual virus in the goat white blood cells (WBCs). It is useful for screening early infections before antibodies develop (a process called seroconversion) as well as catching dormant infections. PCR is very sensitive and specific. Screening with PCR is frequently done on pooled samples. While pooled samples are good for herd screening, there is a drawback. If a positive sample in the pool is missed, the pool will be negative. At this point, all the samples are considered negative and the herd is falsely classified "clean."

The differences in AGID, ELISA, and PCR testing are some of the reasons someone can buy a "negative" animal from a herd, yet have it come up positive months or days after purchase. Other sources of contagion are described in the individual disease sections below.

For any disease testing, it is best to work with a veterinarian, when possible, to make a long-range plan for your herd. Every situation is unique. There is no real one-size-fits-all protocol that covers all tests and all herds. In this edition, I have beefed up the testing section as well as CAE, CL, Johne's, and scrapie sections to reflect the increasing interest in "disease-free" goats. All the testing required to truly qualify for "disease-free" status, as well as the extra handling, record-keeping, and long-range commitment should be reflected in the price of any animal you buy or sell that has that designation.

CAPRINE VENIPUNCTURE

SUPPLIES

Livestock clipper
Alcohol wipes
Vacutainer and holder
Specimen tube (usually red top for serum testing)
18–20-gauge needle
Pens and labels for tube

1. Shave the neck over the jugular vein.
2. Cleanse the area with alcohol.
3. Secure the vein with your thumb below the blood vessel so it stands out. This acts as a sort of tourniquet.
4. Insert needle into the vein at a 45-degree angle. Use a quick, smooth puncture to reduce pain and trauma to the goat.
5. Hold the needle steady while filling the tube.
6. When the tube is full, either fill a second tube or remove the needle from the goat. Most tests require a minimum of 1cc blood for the test. Try to fill the tube completely, but don't discard it if you only get it partially filled.
7. Identify the sample with at least two forms of ID. Useful information for labeling a tube could include the goat's name, the tattoo or ear tag, registration ID, date of birth, or the name of the owner or herd.
8. Prepare for shipment according to the directions provided by your lab.

HOOF TRIMMING

Hooves are the foundation of the goat. The nimble movements and breathtaking climbs that goats love to perform are impossible with overgrown, cracked, or sore feet. Most goat owners dislike the chore of hoof trimming and avoid it whenever possible. But it is one of the nicest things you can do for your goats—even if they don't believe it.

Standard schedules recommend trimming hooves every three months. In reality, the time between trims should be based on each individual goat. Goats that spend time on hard ground and rocks wear their hooves down naturally. When goats are kept on soft bedding, their hooves grow rapidly. Goats that spend time in wet areas collect crud between their toes and in the overgrown part of the hoof, which can lead to foot rot.

The main tool needed for trimming hooves is a pair of good trimming shears or a knife. A hoof pick can be useful for cleaning out crud. Some people "sand" the hooves as a finishing touch using a hoof file, hoof plane, or even a Dremel. Many goat keepers recommend shears because a knife is too difficult to use on tough hooves. After trying several products, I discovered the "magic" orange-handled shears. These Teflon-coated shears, available from livestock supply catalogs, stay sharp and do not fall apart easily. Ergonomically designed, they allow more trimming without causing hand cramping or blisters.

The best way to protect your hands while trimming hooves is to wear gloves. Hooves are hard, and the knife or shears can slip as you trim a less-than-cooperative goat. One goat

This is what an untrimmed kid hoof looks like. *Jen Brown*

The kid in the foreground is disbudded; the other is naturally polled. *Carol Amundson*

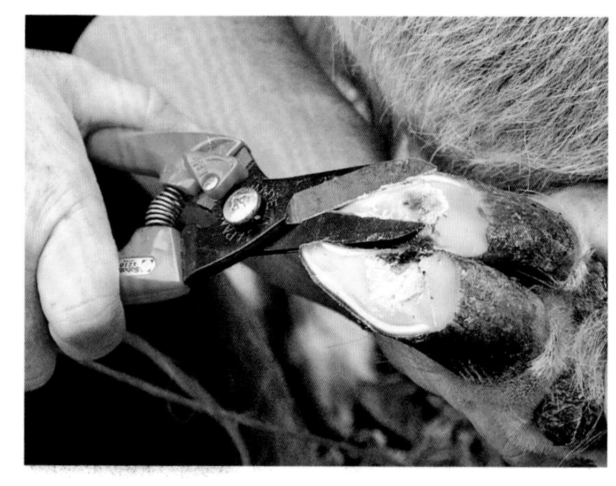

Teflon-coated trimming shears work well for trimming goat hooves. *Jen Brown*

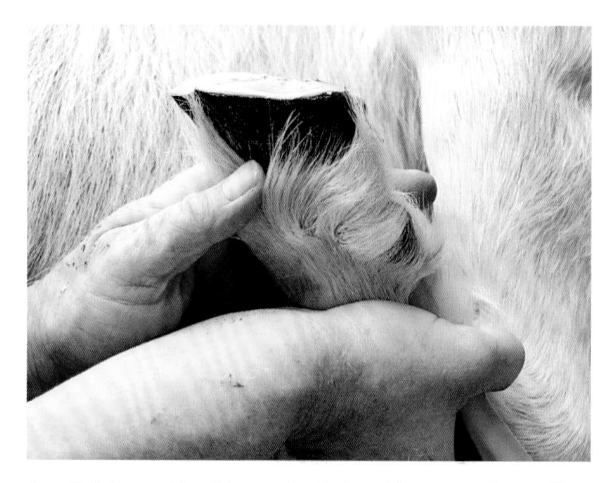

A well-trimmed hoof keeps the feet and legs sound as well as preventing foot rot. *Jen Brown*

owner I know had to visit her local ER after clipping the tendon in her thumb while trimming the hooves on a buck.

HORNS OR NO HORNS

One of the most hotly debated topics on goat forums and social media involves the horns on goats. If you raise or show registered dairy goats, the decision to dehorn is almost a nonissue. The major dairy goat registries require that registered animals be disbudded or dehorned. Pet goats, meat goats, and hair goats are more commonly kept with their horns. While this is a personal issue, I urge you to look carefully at the pros and cons if you are going to breed your goat and later sell the kids.

Most goats grow horns. The exceptions are goats that are naturally hornless, called polled. Polled goats have rounded knobs where horn buds would normally be, and their heads are smooth. This is a recessive genetic trait inherited from both parents. When two horned goats produce a polled kid, it means that each of those goats has one polled gene and one horned gene. Horned genes are dominant, so the goat with only one gene for the trait will have horns. The world's longest goat horns on a living goat measured a tip-to-tip spread of 55.11 inches in February 2018 on Rasputin, a Walliser Black-Necked goat in Austria.

BREEDING POLLED GOATS

It sounds like a good idea to breed polled to polled and never have to deal with horns. Unfortunately, the gene for hermaphroditism is linked to the same gene that produces polled goats. Hermaphroditic goats are neither exclusively male nor female but have intersex features. Sometimes these kids are born with genitals and urinary parts combined or hanging out of the body.

There are groups of breeders—and even a Facebook group—that promote breeding polled to polled, claiming that there is no issue in their herds. Since most of the studies regarding polled to polled breeding causing hermaphrodites were done in dairy goats, breeders of other types of goats sometimes feel the risk is worth it. I have read personal farm accounts of Boer goats producing intersex or hermaphrodite kids when trying polled to polled crosses. I personally saw such kids when a friend who bred Myotonic goats was unaware her buck was polled. She had several abnormal kids that had to be put down because of urinary deformities and genitalia outside of their bodies. Depending on the situation, the odds could be as high as 25 percent of kids being born abnormal. If a polled buck happens to be clean for throwing intersex kids, a breeder would never see the deformity. There is currently no way to test for this.

THE CASE FOR HORNS

On the practical side, horns serve to disperse heat. The large blood vessel in the center of the horn circulates warm blood from the body through the horn. Some of the heat is then let

Horns help circulate blood throughout the body, which becomes an important consideration for range herds in hot, dry climates. *Jen Brown*

Dairy goats are generally disbudded because horns on the milk stand can be dangerous in close quarters. These goats must twist their heads to get into the stand and could injure goats beside them as they swing their heads. *Carol Amundson*

off through the hard shell of the horn, allowing the cooler blood to return to the body. This natural air-conditioning can help goats in hot, dry climates or those that are working hard, such as pack or cart goats.

In open country, horns serve as predator protection. A goat is good at lowering its head and scooping the horns under an offending coyote or dog. If placed in the right spot, a solid hit with a horn can kill or severely wound another animal. This is also a good place to remind goat farmers that people are also classed as animals. There have been people severely injured or killed by goats, most notably a man who mistreated his horned goat. The poor animal rammed him multiple times causing a ruptured stomach and killing him on his front porch.

One unacceptable reason to keep horns is the claim that they make a good handle. In a pinch, I have moved goats by holding onto or pulling on their horns. However, this isn't really a good method. Goats are smart and may be trained to lead with a collar. Pushing or pulling on horns can lead to aggressive behavior in your goat. Messing with a goat by pushing or pulling on the horns or poll is seen as a dominance challenge by the goat. The most natural movements for a goat are to swing the head, ram, or lower the head and hook upwards. In spite of possible hazards, I would probably choose to keep the horns on goats if I kept them on the range or had a pack or cart herd.

THE CASE AGAINST HORNS

I will tell you from the start that I am not a big fan of horns. I have both horned and dehorned animals. It is my opinion that horns are a hazard in most situations—to people and goats!

Animal rights activists argue that goats are mutilated with hot dehorning irons in the name of looks and convenience for the goat owners. In my experience, nothing could be further from the truth. Except on range, horns no longer have a good place on modern farms. Goat horns get caught in fences, trees, and other tight places. Most commonly, the goat gets its head or just horns caught in the fence or a feeder. I have seen photos of goats whose horns have gotten wedged in the crooks of trees, caught in basketball nets, or hanging from aerial wires. I took collars off my young bucks after finding one goat strangling another goat, his horn tangled in the other goat's collar. One of my first Boers was boarding at a veterinarian's farm when she caught her horns in the fence. By the time she was rescued, her struggles had caused her to break her neck.

Horns that get stuck sometimes break off. Because of the large blood vessel running through the center of the horn, the blood loss from this type of injury is substantial, requiring veterinary care. My doe broke her horn off in a fence and bled so severely she was covered with blood. I couldn't stop the hemorrhage. Veterinary repair involved anesthesia before sawing off the remains of the horn. Along with the shock of her trauma in the fence, the poor girl had a bad response to the anesthetic, a common occurrence in goats, particularly Boers. Overall, while the goat survived, her pain and suffering were much worse than if she had simply been disbudded.

A dehorned goat is also easier to handle. Hornless goats generally make better companion animals. Horns catch on clothing or give owners bruises. I can't tell you how many ripped shirts I have gotten from my horned does catching me wrong. Hornless goats are easier to put into a stanchion and don't need to be helped out of the hay feeder because their horns are stuck. I always had to watch for Bridget, one of our horned Boers that got stuck on a regular basis.

There is a very real possibility for horned goats to injure people. Much as we'd like to believe that all goats are raised properly, some learn to butt people. This behavior can be painful and annoying from a goat with no horns. From a horned goat, butting can injure or kill someone. A swinging goat horn is just the right height to accidentally hit a child in the eye.

Use caution when mixing horned and dehorned goats in the same herd. Make sure the horned animals aren't too aggressive. Goats naturally work out a hierarchy or pecking order. They can be brutal about using their horns on other goats to establish their position in the herd.

HORN PREVENTION OR REMOVAL

TECHNIQUE	COMMENTS
Disbudding	Recommended method for caprine horn prevention Perform when kid is 3 to 7 days old or as early as horn buds are found Causes measurable stress seen in increased cortisol levels Take care not to overheat the brain Give tetanus antitoxin when immunity is questionable Scurs or horn growth possible if cautery is incomplete Injury or death possible if overheated
Banding	Removes small to medium horns Flies can be a problem during fly season Pain may occur when nerve is severed Scurs may replace horn when bud isn't completely destroyed Sometimes bloody Requires a disbudding iron or heated rod for cauterizing bleeds Works on some goats, unsuccessful on others
Veterinary removal	Costly Difficult to find vets experienced in horn removal Vet can provide pain blocks or anesthesia during procedure
Sawing	Cauterize bleeds using a disbudding iron or heated rod Scurs may replace horn when bud isn't completely destroyed Traumatic and painful to the goat Saw, especially wire blades, can slip and cause injury to you
Caustic paste	Failure is likely Hazardous to goats' eyes Kids rub or lick paste off pen mates, causing burns More commonly used on cattle; not recommended for goats
Gouging	Bloody Device is used to scoop out the horn buds More commonly used on cattle; not recommended for goats
Clove oil (eugenol)	Experimental—newer method of chemical disbudding Cortisol and stress level in kids comparable to cauterization No open wound Faster healing than cautery Possible injury or death due to injecting incorrectly Scurs or horn growth possible if cautery is incomplete

DISBUDDING

The best way to remove horns is to catch them before they develop. Large buds and small horns that have already formed take longer to disbud, which is more traumatic for both the goat and the handler. When performed too late or too lightly, disbudding fails to completely destroy the horn bud. Incomplete horn removal results in scurs as the animal ages. Scurs are deformed horns. Scurs are more common in bucks since male hormones strongly influence horn growth. Some bucklings are born with tiny horn buds already present. By comparison, does show slower growth with less pronounced horn development.

Disbudding is one of the hardest tasks for the new goat owner to learn, but it is crucial in most herds. I suggest that you find a goat owner who is willing to teach you the process the first few times. I work with new goat owners all the time and consider it a duty to pass along the skill I learned from the person who sold me my first animal. Improper disbudding technique has resulted in injury or death of kids, even when performed by a veterinarian.

In the United States, not many veterinarians perform disbudding. Goats are considered a minor species, so veterinarians schooled in the United States get limited training specifically in goats. Additionally, goat owners in the United States commonly perform disbudding as a routine chore. I have seen some pretty unfortunate looking disbuddings by veterinarians, who have to learn by practice just like everyone else. However, in other parts of the world, such as the United Kingdom, disbudding is illegal except when performed under anesthetic by a veterinarian.

Recent studies show disbudding increases stress and causes an increase of the stress hormone cortisol during disbudding. The younger the goat and the smaller they are, the better. When disbudding is performed in very young kids, they become stressed, but the process only lasts 15 to 30 seconds. The kid recovers quickly and returns to mom or a bottle to nurse.

Because of the smaller size of goats compared to other ruminants, they are known to be quite sensitive to standard anesthetics, including lidocaine. There is a possibility of toxicity when applying analgesia or anesthesia by inhalation or injection. Some owners use children's ibuprofen or a 325-milligram dose of aspirin 30 minutes before the procedure. There isn't often bleeding with a proper cautery. However, it is useful to remember that aspirin can inhibit blood from clotting. An herbal called Ow-Eze from Molly's Herbals is also popular for pain relief.

The best age to disbud is between three and seven days old since horn buds grow very quickly. There is a much larger growth area at the base of the horn to be destroyed in kids compared to horn buds in calves. The new kid must be strong on its feet and have a palpable horn bud. The timing is subjective; some kids (especially doelings and miniature breeds) are slow to develop their horn buds. Some bucklings are born with the buds already starting to grow. The earliest you can feel the bud after the kid is strongly on its feet is the best time to disbud.

Disbudding is done with a hot iron specially designed for dehorning goat kids. (I use the Rhinehart X-50.)

In recent years, I have been trying a variety of disbudding irons. I am still most happy with the Rhinehart X-50 because it has been the most reliable. The biggest disadvantage to the Rhinehart is the cord. The cord requires disbudding to be performed near an outlet. I have also had to replace the cord on my iron because it got melted and burned by the iron.

I was very excited when I purchased the Steribud cordless battery-operated iron. The first season, I was able to use it out in pasture on my kids that were dam raised. It was really nice not to deal with a cord and the kids ran right back to mom when they were done. I also found it convenient to use on kids that were brought over by other breeders needing disbudding services. I could disbud right next to their vehicle. Unfortunately, the battery is very sensitive if overcharged. Within a year, the iron started to be cooler and have a weaker and shorter charge. Finally, it quit altogether. Given that it was more than twice as expensive as the Rhinehart, I have gone back to my standard iron.

Another iron that I have tried but don't prefer is gas operated. The Portasol operates with propane and seems to be a good cordless option. Unfortunately, I have had some difficulty getting it charged with propane or keeping it heated.

Whatever iron you choose, first, use a hair clipper to remove hair around the bud. Trimming prevents the strong smell of burning hair and helps you see very small horn buds.

This step is especially helpful to new goat owners. In larger operations, hair clipping is

DISBUDDING

REQUIRED
Disbudding iron
Wire brush

OPTIONAL
Tetanus antitoxin
Towel or puppy pad
Disbudding box,
homemade or purchased
Small hair clipper
Gloves
Solarcaine or other burn relief spray

1. Preheat your disbudding iron. Be sure to avoid placing it in a spot where it may start a fire or where the cord will be tripped on by goats or people. This tool gets VERY hot!
2. While waiting for the iron to heat, give the dose of tetanus antitoxin (unless the kid already has immunity from its dam or an earlier injection).
3. Restrain the kid in a box or on your lap (covered with a towel or puppy pad if desired).
4. Clip the hair from over the buds.
5. Check to see that your iron is hot enough. The tip will glow red if you are in a darkened room. The tip should leave an immediate round burn scar on a piece of wood.
6. Set the tip of the hot iron directly over the center of the horn bud.
7. Apply gentle, even pressure while twisting the iron back and forth in a circular motion to help provide an even burn. Some people count to ten slowly or repeat the phrase "This will save your life" ten times.
8. Raise the tip of the iron to see if the burn is complete.
9. Remove the burnt skin, hair, and ends of the loose horn (especially important for bigger buds on older kids).
10. Reburn as necessary until there is a clean, copper-colored ring around the bud.
11. Repeat on the other bud.
12. Spray head with burn relief as desired.
13. Give the kid a little extra petting or a bottle and release.

Late or improper disbudding can result in deformed horn growth called scurs. *Jen Brown*

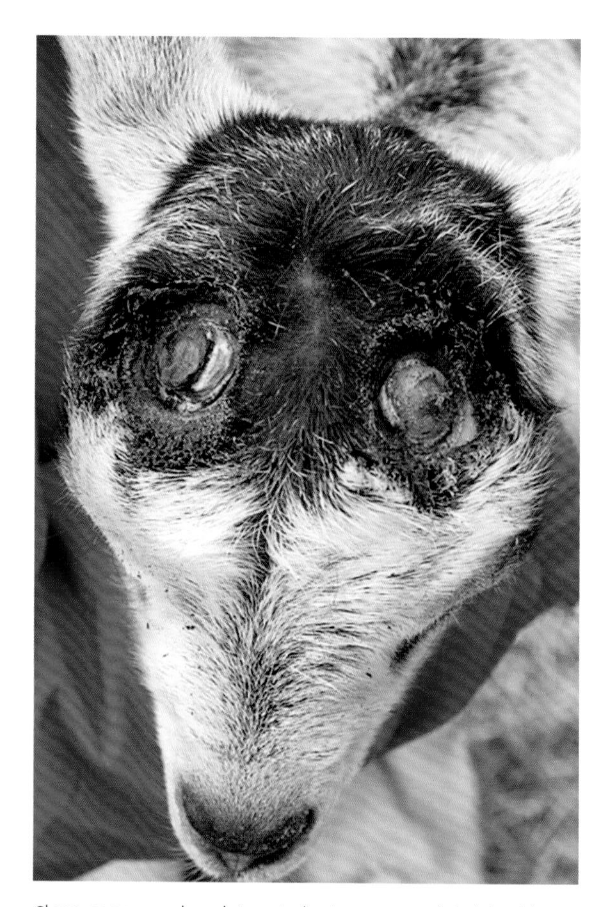

Clean copper-colored rings indicate a successful disbudding. *Jen Brown*

typically considered unnecessary and time-consuming. An inexpensive wire brush from the hardware store cleans the excess hair and other burned debris from the tip of the iron to make it more efficient. While I am guilty of neglecting to use gloves, it is best to protect your hands with work gloves to prevent burns.

A disbudding iron is very hot and burns at a simple touch. (I have the scars to prove it!) A disbudding box is a simple structure used to restrain the kid during bud removal. Oftentimes, it's just as easy to disbud without the box. I found a disbudding box useful in the beginning or for larger, unruly kids. However, I haven't used one for years.

A firm hold with the kid over my knees gives me sufficient control. Bucklings have been known to pee when being burned. Place a puppy pad or towel on your knees under the kid to keep dry. I have seen some breeders disbud while straddling a towel wrapped kid and holding its head pinned to the ground. A firm grip is necessary when working without a box.

If the doe was vaccinated for tetanus before kidding, the kid will have acquired immunity from its mother. If the vaccination status of the dam is unknown, give the kid a dose of tetanus antitoxin at the time of disbudding.

Some breeders use topical antiseptic sprays or numbing sprays during disbudding to stave off infection and provide comfort. Solarcaine or other sunburn sprays seem to provide relief, if only for the goat owner. Choose a product with the highest level of lidocaine and apply immediately after disbudding to cool down the head. It isn't helpful to apply spray before the procedure, as the numbing effects are minimal, and the spray is burned off in the process of disbudding.

The best way to eliminate horns in your herd is to disbud kids within the first week of life. *Shutterstock*

The initial chance of infection from disbudding is slight because the hot iron seals the bud. Greater chances of infection arise later, when the head starts to heal and the scabs get knocked off, which can leave a slightly bloody wound. A wound spray can also help, either at the time of disbudding or when there is bleeding from the scab. Be aware that some of these sprays are not allowed for use in food animals.

While disbudding, it is okay to go back and forth from one bud to the other to allow the spot to cool a bit. The temperatures of individual disbudding irons vary greatly. Know your iron. Remember that you only want to burn enough to remove the bud; too much burning will overheat the kid's brain or break through the skull. Be aware of the kid at all times when disbudding rather than relying strictly on a time count for the process. Do not use excessive pressure on the head with the iron.

Older irons wear out, so watch the condition of your equipment. One herd manager who had disbudded hundreds of kids without incident suddenly had a number die after disbudding. After examining the kids and the iron, he discovered that the iron had worn thin and the hot, sharp metal was piercing the skull, causing severe brain injury. This is an extremely rare occurrence but bears mentioning, especially if you purchase used equipment.

An adult goat can be dehorned by banding or sawing, but this is traumatic to the goat and increases the risk of infection. *Jen Brown (top), Carol Amundson (bottom)*

CLOVE OIL DISBUDDING

Pure oil of clove bud contains a high percentage of eugenol, a substance known to destroy tissue. In recent years there have been several published studies plus individual farm tests using this oil to kill horn buds to prevent horn growth. A 2014 study in Iran showed 100 percent success with using clove oil injected into the horn bud. Later studies had a variety of results, but mostly showed clove oil to be as effective as cautery disbudding. Until more people have tried it, this technique remains rather experimental. Clove oil is widely used in dentistry as a numbing agent and antiseptic.

The studies had best luck with a measured dose of 0.2cc oil of clove with at least 80 percent eugenol for full-sized kids and 0.1cc for miniatures. It is important to inject the oil in one injection from the back of the head straight into the bud. The needle should be parallel to the top of the skull. Injecting straight down into the bud had serious consequences if the oil ended up deeper than the horn bud in the skull. In worst cases it killed the kid. Care should also be taken not to inject into a blood vessel.

I am going to do more research but would like to try this method of disbudding. One safety tip—wear eye protection. Clove oil can cause serious injury when splashed into eyes. There's a Facebook group available for people who are interested in disbudding via clove oil.

LIVESTOCK GUARDIANS

Goats can be the victims of accident, theft, or predation. A livestock guardian (LG) is a good investment, particularly when the goats reside a distance from the main farm. Most commonly, the guardian is a dog, but llamas and donkeys are also used. While not a cure-all, the presence of a guardian can deter predators. The true guardian that runs with the herd is a special animal. Personality and training both factor into its effectiveness. Look for breeders who raise guardian animals. Remember that even good livestock guardians need to be trained to respect the stock. Before getting a livestock guardian of any kind, it is important to do your research and take as much care selecting your guardian as you do selecting your goats.

Livestock guardian dogs (LGDs) have been used for centuries in Europe. The common breeds originated there, often carrying the names of their region, including Great Pyrenees, Akbasch, Maremma, Anatolian

Livestock guardian dogs require strict training to be effective. They must respect the goat herd. *Carol Amundson*

Shepherd, and Komondor. Some, like the Great Pyrenees, prefer a large territory.

Guardians need to be capable of independent thinking. Overly shy animals should be avoided. Livestock guardians are different from herding dogs in that they do not have the instinct to chase. While younger dogs may do some chasing, the older dog is more sedentary, preferring to hang out with the herd. You do not want a dog that is aggressive toward people or prone to wandering.

You may find an LGD that is a crossbred animal of several guardian breeds. A dog from working parents will allow you to assess the pup's potential as a working LGD. Some farms raise a mix of guardian breeds and have very good dogs. My best LGD was a mix of Sarplaninac, Great Pyrenees, and Akbasch.

Do not get a cross of a herding dog and a guardian dog in the belief that this cross will be able to do both functions. Livestock guardian breeds have a very low prey drive and have been bred for many years to be independent. Herd dogs are almost the opposite—very high prey drive and an inborn desire to follow the orders of its master.

Training depends on your needs and the nature of your farm. Animals trained to range are frequently restricted from human contact at eight to twelve weeks of age to facilitate bonding with the goats. Dogs on smaller farms should be more socialized to people. Many of these large dogs are very calm and seem almost lazy as they lie about in the pasture during the day. At night, they go into action, patrolling boundaries and barking warnings. This barking may be a problem for small goat farms that are in more residential areas with close neighbors.

Expect to conduct extensive training of your dog, including teaching it not to chase or roughhouse with the goats. These dogs mature slowly and do not leave puppyhood before the age of two or three years. Immature dogs should be kept in smaller areas and monitored. Exuberant young animals have been known to inadvertently maim or kill the animals they are supposed to protect.

Unfortunately, the large-breed canines are not long-lived animals, often having an expected lifespan of eight to ten years. Consider adopting a younger dog as the old guardian ages so that the youngster may be trained by a canine mentor.

Llamas and donkeys use many of the same feeds and medications as goats, which make them easy to manage as livestock guardians. Both species have an instinctive dislike of canines. They will drive off intruders by calling out, kicking, and chasing the stranger. Intact males are not usually used due to their aggression toward goats, including a tendency to mount or attempt breeding. Most often, a single llama or donkey is used in the herd since, like a dog, the guardian needs to bond with the herd, not with people or others of their species.

NEUTERING UNWANTED MALES

Neutering isn't a pleasant task. However, it eliminates the characteristics that make bucks objectionable, such as odor and breeding behaviors. Wethers make excellent companions or working goats. In meat herds, wethers reduce rut behaviors. Fiber-goat wethers have cleaner fiber. Neutering can also extend the goat's lifespan, since wethers do not go through the stress of rut every fall, which tends to make bucks shorter-lived on average compared to other goats.

Bucks are usually born with their testicles descended. While neutering is possible even at a few days of age, some breeders choose not to castrate until the animal is eight to twelve weeks old. This allows the urethra time to develop and may help prevent urinary calculi problems later—although other reports indicate that the age at neutering is a less important factor than the animal's diet.

Ethical and market considerations should be taken into consideration when deciding on your farm's neutering philosophy. In meat animals, some buyers require intact animals, while others prefer wethers. Some sellers do not allow any nonbreeding male to leave the farm without neutering. This keeps unsuitable genetics from entering the gene pool as well as preventing new owners from unwittingly purchasing a pet goat that later turns into a stinky, unpleasant menace.

Methods of neutering include cutting the scrotal sac and removing the testicles, crushing or "crimping" the spermatic cord with an instrument known as an emasculator (Burdizzo), and banding the testicles with an elastrator band.

On our farm, we typically castrate by cutting the male kids at a young age when their testicles are still small. Banded kids seem to be in distress longer than those that are cut or crimped. Some European countries consider banding inhumane and have made it illegal. Because the scrotal sac actually rots off, this technique has the highest risk of infection.

Some breeders give pain medication prior to performing castration or other traumatic procedures. Oral pain meds are poorly absorbed by goats, but a regular aspirin given at the dose of one tablet per 10 pounds or oral Banamine at 1 milligram per pound may be given a few hours before the procedure. An injection of Banamine can be procured from a vet and given a half hour before cutting or crimping. Fias Co Farm in Tennessee sells an herbal painkiller known as Ow-eze on its website. Some other breeders use willow branches for the natural salicylates that act similarly to aspirin. Be aware that aspirin can reduce blood clotting and might lead to extra bleeding after surgical castration.

Most large herd operations castrate their own male kids, but veterinary help may be preferable for specific cases or smaller herds. The vet has

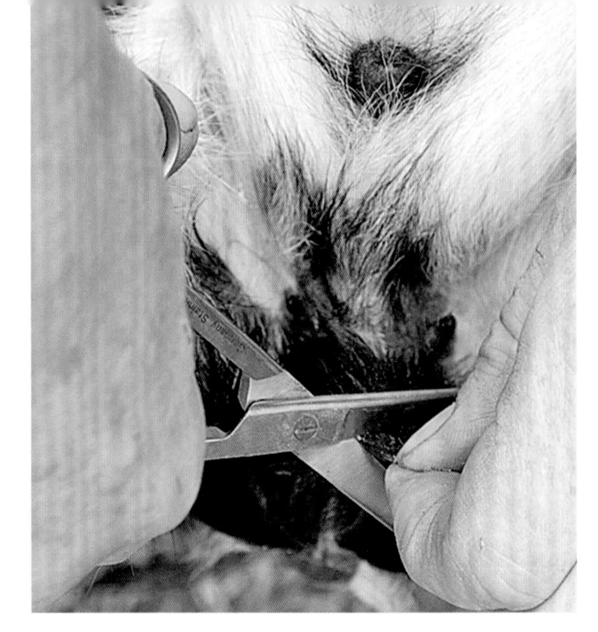

Cutting the scrotal sac and removing the testicles is another method of neutering. This method is still unpleasant but less traumatizing to the goat. *Jen Brown*

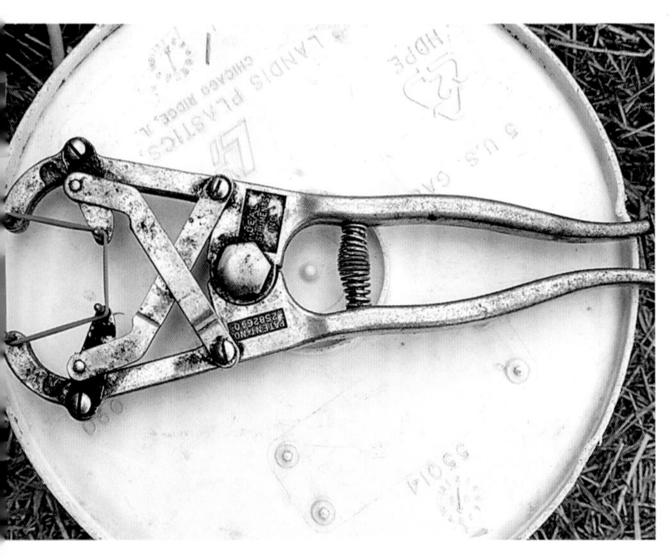

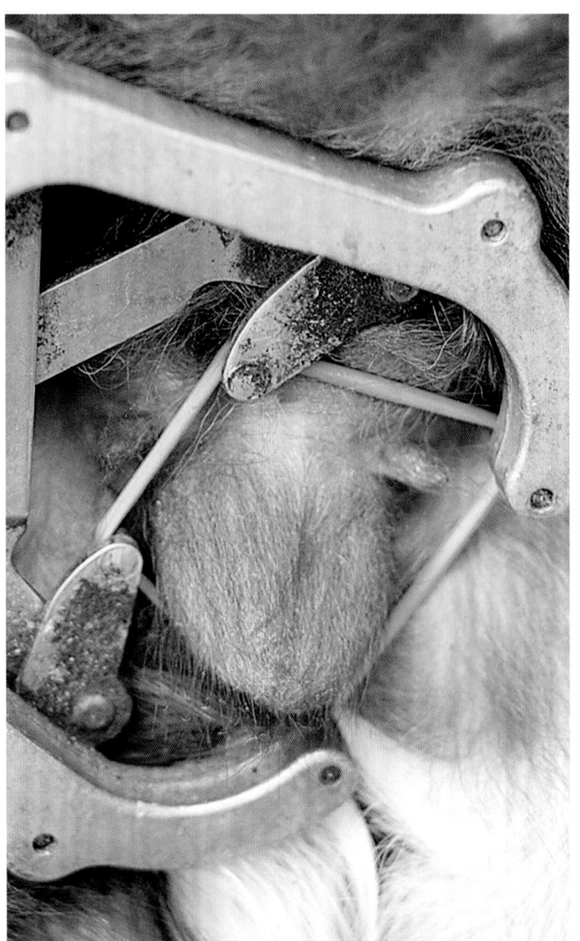

Above and right: Banding the testicles is one method of neutering a buckling. This method leaves the goat vulnerable to infection and increases the level of trauma to the goat. *Jen Brown*

NEUTERING

METHOD	PROS	CONS	SUPPLIES	SAFE AGE
Banding	Inexpensive Bloodless	Least humane method Technique error may leave a testicle and fail to sterilize Risk of tetanus	Elastrator Bands or rings Tetanus antitoxin if not vaccinated	After the testicles descend More traumatic the larger the testicles
Cutting	Inexpensive Most reliable method	Open wound Risk of tetanus Bloody, may be disturbing Risk of excess bleeding in older kids	Surgical scissors or scalpel Disinfectant Tetanus antitoxin if not vaccinated Pain medication if desired	After the testicles descend More traumatic the larger the testicles Do not cut kids with a scrotal hernia Bucks over 6–8 weeks should be done by a vet with anesthesia
Emasculation	Quick recovery No cutting or blood involved No chance of infection Relatively humane	High initial equipment cost Technique error may result in incomplete castration	Burdizzo or Nipper Pain medication if desired	4 weeks or older May be used on full-sized bucks
Vasectomy	Useful for teaser buck Very reliable	Expensive Leaves breeding traits intact	N1A	Any age

sedation available. Full-size bucks may be neutered by crushing the spermatic cord with a Burdizzo.

We have had this done by our vet rather than attempting it ourselves on a full-size buck. Veterinarians may also perform a vasectomy if you want a teaser buck for the herd. This method does not eliminate breeding behaviors, so it's unsuitable for pet goats.

Always begin any home neutering operation by washing your hands and the instruments thoroughly with soap and water. Also wash the buckling's scrotum, if it is dirty. You will need a helper to cradle the kid on the helper's lap, with the kid's legs held firmly apart and out of the way in either hand.

NEUTERING BY CUTTING

You'll need these supplies:
Surgical vaccine

1. Clean your hands thoroughly and use surgical gloves if available.
2. Disinfect the scrotum with iodine or Nolvasan.
3. Dip scissors or scalpel in disinfectant.
4. Push the testes to the upper part of the scrotal sac.
5. Cut off the lower one-third of the scrotum using a surgical scissors or scalpel to expose the testes.
6. Using your fingers, grasp one testicle firmly. It will be slippery, so try to avoid letting go or it will retract into the scrotum.
7. In young kids (less than four weeks), pull down firmly and steadily until the cord breaks. In older kids or adults, scrape the cord with a knife or scalpel until it breaks. Repeat with the other testicle.
8. With the kid still held belly-up, pour disinfectant into the wound, making sure the cord ends are retracted or scraping off the excess length to prevent infection.
9. Give a shot of tetanus antitoxin if the dam was not vaccinated prior to kidding or if the kid's immune status is questionable.
10. Do not excite a kid immediately after castration. Allow him to nurse mom or a bottle.

NEUTERING WITH AN ELASTRATOR

You'll need these supplies:
Iodine or Nolvasan
Elastrator and bands
Tetanus vaccine

1. Disinfect the scrotum with iodine or Nolvasan.
2. Dip the band and elastrator in disinfectant.
3. Load the elastrator with a band and lock it open with the band stretched.

Kids seem to recover very quickly from brief surgical procedures, such as neutering, usually wanting to nurse or take a bottle after being let go. *Jen Brown*

4. Gently maneuver the testicles into the lower part of the scrotal sac.
5. Position the band around the testicles, with both testes below the band and neither teat within the band.
6. Release the holding clamp and roll the band off the pins and onto the scrotum.
7. Reapply disinfectant to the band.
8. Give a shot of tetanus antitoxin if appropriate.
9. In the following weeks, monitor the castration site for infection or complications while the testicles atrophy.

NEUTERING WITH AN EMASCULATOR

You'll need these supplies:
Emasculator
Pain salve (such as Burn Jel)
Iodine or Nolvasan

1. Grasping the scrotum in one hand, gently move the testes down into the lower part of the scrotal sac.
2. Locate the spermatic cord on one side and use your fingers to position it as far to the outside of the sac as you can.
3. Position the crimping jaws of the emasculator so that as little of the scrotum as possible is between the jaws and the cord is resting next to the tooth of the tool.
4. Be aware of the location of teats. Do not crimp those by accident.
5. Squeeze the implement completely closed until you hear it click to crush the cord.
6. Leave the clamp in place for a count of five.
7. Open the emasculator and repeat the process for the other side.
8. Pain-relieving salve may be applied to the pinch. I use Water Jel's Burn Jel, which has 2 percent lidocaine content compared to 0.5 percent for most over-the-counter products, including Solarcaine.
9. In the following weeks, monitor the scrotum to be certain that the crimping has been effective. If after three to four weeks either testicle has grown, or one is larger than the other, repeat the process on the growing testicle.

Emasculators

There are several brands and models of crimper on the market, the most well-known being the Burdizzo. These tools are made for larger animals such as cattle, so the smaller size-9 model is recommended for goats. Some emasculators do not close completely but leave a small gap. This design is difficult to use on goats since the spermatic cord can skip into the gap and not be properly pinched. The scrotal sac remains intact after using an emasculator, so goats wethered in this manner will always have an empty sac—a small reminder of the procedure.

PARASITE PREVENTION

Every goat keeper needs a parasite control program. Talk to local vets and experienced owners about this subject, since there are regional differences in the types of parasites and effectiveness of treatments. Most often, parasites develop due to contamination in feeding areas.

Parasites are responsible for a large percentage of losses accounted for by poor feed utilization, which leads to increased feed costs, as well as decreased production and unthrifty animals. More severe cases of parasite loads increase mortality rates.

Parasites are organisms that live either on or in your goat. They take food and shelter from the animal but give nothing in return. Some of these creatures are species specific. Others are less discriminating, including ringworm, toxoplasmosis, and tapeworms, which will infest multiple species—even humans. The most common caprine parasites are lungworms, barber's pole worms, and liver flukes, in addition to coccidia, but a host of other critters can infect goats.

Stressors, such as weather changes, traveling, or new herd mates, can trigger an outbreak of parasites that have been dormant. Warm, moist conditions are the prime setting for parasite problems to develop.

GUIDELINES FOR PARASITIC PREVENTION

- Avoid contamination of water buckets by animal or bird droppings.
- Place feed off the ground to keep feet and manure out of the feed sources.
- Maintain low stocking rates: no more than six to eight small ruminants per acre. Alternate grazing of cattle or horses can help.
- Rotate grazing areas and pens.
- Ensure proper trace elements in diet through Bo-Se injections, copper boluses, and mineral supplements as needed to keep up natural immune function.
- Sanitize kid-rearing pens and equipment between batches of kids.
- Select medication and dosages based on herd observation and worm load testing.
- Treat with drugs based on need or FAMACHA scores, dosing only goats with high parasite loads.
- Withhold feed for twelve hours prior to administering oral wormers.

External parasites, such as ringworm, lice, or mites, make animals uncomfortable and damage the skin and hair coat. Sucking lice, when present in large numbers, can cause anemia. External pests are generally no more than a nuisance, except when their numbers are high or in hair- and hide-producing herds. Sucking lice also can be extremely damaging to goats that have another condition already compromising their health. Some of the same wormers used for internal parasites are useful for controlling sucking insects. In other cases, a medicated shampoo or powder may be necessary.

When talking parasite control, goat owners most often mean internal parasites. Coccidia are microscopic organisms. Worms such as the barber's pole or tapeworm are large enough to be seen with the naked eye. These creatures live in the caprine digestive tract, taking nutrients from the intestinal contents or sucking blood from the animal. More difficult organisms to treat are those that live outside the intestines and do not show up during fecal exams. Liver flukes and lungworm are examples, as is the deer worm, which attacks brain and nervous tissue.

As with other management tasks, approaches to parasite control range from nonintervention to regular testing and treatment, as well as automatic, timed treatment without testing. Unfortunately, some wormers are becoming ineffective as parasites build up resistance. Many wormers now reduce parasite loads by fewer than 5 percent after treatment. The old practice of worming herds of goats on a schedule that doesn't consider the presence of parasites is now being discouraged as much as possible. Incomplete treatment, or treatment with the wrong medicine, helps develop resistant parasites. Overused drugs then become useless in the case of actual infection.

A fascinating side note: Australian researchers, working to combat barber's pole worm in sheep, have a promising new parasite treatment in development using spider venom. An assortment of different spider venoms are proving effective against a wide variety of parasites.

Clinical signs of parasitism include rough hair coat, diarrhea or soft feces, swelling under the jaw, and pale membranes in the inner eyelids. The best first step in establishing effective parasite control is to test your goats. Identifying a baseline of pest levels will help you determine which wormer will be most effective. Owners can take individual goat fecal samples or a mixed herd sample to their vet for a worm count. There are kits for owners who have access to a microscope and the skill to perform their own fecal examinations. Check online for instructions on performing your own parasite testing.

First used by meat-goat farmers in the United States and now universally in use, the FAMACHA parasite control system was developed in South Africa. This method measures anemia, useful in detecting the barber's pole worm, *Haemonchus contortus*. This method must be used in hot weather, when parasites are active.

FAMACHA

The FAMACHA system, developed in South Africa, was brought to the United States by the American Consortium for Small Ruminant Parasite Control (ACSRPC). The ASRPC website is a wealth of useful parasite identification, treatment, and control information for goat and sheep owners. The FAMACHA system allows small ruminant producers to estimate the level of anemia in their goats and, in turn, determine the presence or level of barber's pole worm in the goat or the herd.

The barber's pole worm is the most common cause of anemia in goats. This parasite has a small "tooth" that cuts into the goat's abomasum (stomach) wall. It then feeds on blood from the wound. The resulting loss of blood creates unthriftiness, weakness, and even death of the host animal.

FAMACHA uses a laminated card showing five color categories corresponding to different levels of anemia. The examiner matches the color of the mucous membranes around the eye of the goat to the chart. Category 1 is "not anemic" while category 5 represents "severely anemic." The card is available only to veterinarians or to individuals who have been trained in the system. Training involves several hours of instruction, as well as demonstrated skill as judged by a certified trainer. It is available online or you could contact your local extension office, so it can help you find someone teaching in your state,

COCCIDIOSIS

Coccidiosis is caused by microscopic single-celled protozoa. Coccidia are more of an issue

CRITERIA FOR CULLING

- Bad bite, broken or missing teeth, problems eating
- Bad teats (too big or too small, more than two, split or double teats)
- Bad testicles (split, too small, infected)
- Bad udders (lopsided, poorly attached), especially in dairy goats
- Does that have not settled for two seasons
- Evidence of abscesses or other disease
- Poor body condition due to illness or old age
- Poor-quality fiber (on fiber goats)
- Structural defects (bad feet, legs, back)
- Poor temperament—overly aggressive or too timid, escape artist

in kids, younger animals, or those with poor immune systems. Older goats usually have resistance to coccidia, which can be present in the gut without causing illness. The species *Eimeria* accounts for most of the coccidia infecting goats. Coccidia infecting other animals or birds generally can't cross-infect other species. Two exceptions are *cryptosporidium* and toxoplasmosis, which can infect multiple species including humans.

Vigilance must be exercised in year-round care of your goats. Their wily nature often exposes them to all manner of health problems.

Because adult goats serve as reservoirs for infection and contaminate the environment with their feces, kids are the most at risk for this illness. Coccidiosis is most commonly seen between the age of four weeks and six months. Built-up manure in confinement settings is frequently to blame for infections. However, infection in pasture also occurs, especially from build-up around waterers, feeders, or housing. Coccidia is one of the leading causes of "poor-doers" in a herd since kids that have had coccidiosis will often have damaged guts and stunted growth.

Prevention is the best medicine for coccidia. Good hygiene and sanitation practices will reduce outbreaks. Clean, dry pens and limiting or preventing fecal contamination of feeders and waterers will reduce the chances of infections. Be certain kids get enough colostrum at birth. Stress, improper nutrition, changes in weather, or other illness will all increase a goat's vulnerability to coccidia.

Treatment methods include amprolium (CoRid), tetracyclines, and sulfa drugs. Kids in larger herds are often treated automatically. Over time, the goats develop resistance to these pests. Coccidiostats for kid-rearing are some of the most frequent wormers given on an automatic basis. Unfortunately, many of these treatments are ineffective due to resistance to the drug.

Veterinarians and breeders in your area can tell you which medications are most effective in your area. Most coccidia treatments used for goats are extra-label. According to FDA regulations, even drugs that can be purchased over the counter require veterinary approval before being used in animals that are not listed on the label as approved. Two very effective medications for coccidiosis, Toltrazuril and Diclazuril, are available in the Unites States on a limited basis. These drugs should be discussed with your vet if they are recommended in your area.

When medicating a group of animals in their drinking water, be sure to offer medicated water as the only source for drinking. Some breeders add drink flavorings or jello to cover bitter flavors and induce the goats to drink. Because some goats don't like the taste of medicine, severely ill animals should be drenched to assure they get enough treatment.

A sick goat is often listless and uninterested in her surroundings. You can tell just by looking at her that this goat is ill. *Carol Amundson*

VACCINATION CONSIDERATIONS

PREFERRED MANAGEMENT STYLES:
Conventional Sustainable Organic

- Risk tolerance for disease
- Known risk of exposure versus risk of adverse reactions
- Effectiveness and safety of vaccine
- Benefit of vaccine versus expense
- Necessity of yearly booster

Medicated feeds containing coccidiostats are used to lower the rate of coccidia shedding into the environment. Deccox (deccoquinate) and Rumensin (monensin) are FDA-approved for goats. Bovatec (lasolid) is approved for sheep, while also being used off-label for goats. Caution is advised, as these need to be properly mixed or may be toxic to the animals being treated. Rumensin is known to be toxic to equines and dogs.

Coccidiostats are considered "natural" and have no withdrawal periods. For a more organic approach, oregano and cinnamon oils are being tested as natural antibiotics. Various herbal mixes that act as much as immunity boosters as antimicrobials are also sold.

VACCINATIONS

Vaccine management differs between conventional and holistic approaches and can be a matter of debate. Vaccines are not harmless, so most goat keepers vaccinate only if the threat is real.

There are two types of immunity. Passive immunity, which is transmitted from the dam, creates healthy kids. Maternal antibodies pass through the placenta to kids in utero. Colostrum provides antibodies that protect kids for the first two months of life. Active immunity occurs after two months of age when an animal is exposed to disease by direct contact or through vaccination.

ARE VACCINATIONS NECESSARY—AND WHICH ONES?

Most goat owners follow some type of vaccination program. Some vaccines are given only when the disease is present in the herd and after management changes have failed. Keys to successful vaccination include proper storage and handling of the vaccine and care to follow recommended inoculation times and doses.

Advocates feel that vaccinations are a form of cheap insurance against disease. The most commonly recommended caprine vaccine is CD/T, which provides three-way immunity against clostridium types C and D (which cause enterotoxemia) and tetanus. There are seven- and eight-way vaccines available, but most are unnecessary for goats.

Compared to human vaccine schedules, our pets and livestock receive boosters considerably more often than we do. Vaccine detractors feel that some (or all) vaccines are unnecessary and that yearly booster shots are too frequent. Some experts believe that annual revaccination is unnecessary because the initial series of shots provides long-term immunity.

CAPRINE VITAL SIGNS

VITAL SIGN	NORMAL RANGE	NOTES
Heart rate	70–80 beats/minute	
Temperature	101.5–104°F	Varies with environmental temperature and activity Lower in the morning Test healthy goats in herd for a benchmark
Respiration	12–15 breaths/minute	
Rumen motility	1–4 movements/minute	Faster after a meal Slower if stomach is empty
Rumen pH	5.5–7.0	

TIPS: HANDLING VACCINATIONS AND OTHER MEDICATIONS

- Store vaccines and medications at the recommended temperature—usually refrigerated.
- Remove the vaccine from the vial only with a sterile needle to avoid contamination.
- Wipe the top of the vial before entering it with a needle.
- Use a separate needle for each animal.
- Do not inject a wet animal—this can increase the chance of an injection site abscess.
- Subcutaneous (SQ) injections are usually recommended for all drugs in goats.
- Keep a record of each animal with the date, type, and amount of injection.
- Follow meat or milk withdrawals from the product insert.
- Avoid mixing vaccines and other drugs; use separate syringes for multiple medications.
- Destroy all needles in a safe manner.
- Keep epinephrine on hand in case of allergic reaction (anaphylaxis).
  Dose: 0.5 to 1.0 milliliter/100 pounds of the 1:1000 product

Measuring rumen motility is a useful skill for detecting abnormalities in your goat's digestion. *Carol Amundson*

The general recommendation for CD/T is to booster yearly, with pregnant does getting their booster four to six weeks prior to delivery. This booster maximizes the antibodies in the dam's colostrum to give kids the highest level of maternal antibodies from their first day nursing or being bottle-fed. Kids who get poor or no colostrum immediately after birth start out at a serious disadvantage since their immune systems need time to develop. Colostral antibodies interfere with immunity being developed from a vaccine. That is the reason CD/T vaccines are not normally given until the kid is six to eight weeks old.

Some breeders vaccinate kids at three to four days old and follow with a booster at three to six weeks of age. This becomes more important if the kid has not gotten colostrum. The other recommendation for kids with less immunity is to wait and vaccinate at one to three weeks. This allows for some limited protection if there is a disease outbreak or the previous dam vaccine failed. Of course, it is also suggested that kids without proper immunity should have a dose of tetanus antitoxin before disbudding or castrating, especially if the dam wasn't boostered during pregnancy.

Rabies Vaccination

The risk of rabies in domestic goats is slight. No rabies vaccine has been licensed for goats. It is a good idea to vaccinate all dogs and cats on the farm for rabies. There is a sheep vaccine for rabies called Imrab. Past studies have shown that this may be effective in goats as well. Consult your veterinarian to decide if it is necessary for your goats.

Other Vaccinations

There are many other vaccines on the market for diseases such as bluetongue, chlamydiosis, E. coli, foot rot, Pasteurella, and sore mouth. Some of these diseases are regionally troublesome, so local practitioners know best whether there is a need for vaccination. Check with a veterinarian to be sure the disease and organism is found in goats before challenging the caprine system with an unnecessary shot.

It is useful to get the vaccination history when you buy a new goat. If you plan on continuing a vaccination schedule, you know which vaccines have been given and when the animal is due for a booster. If the kid you are buying has been given a soremouth (or orf) vaccine, do not bring it to your farm, or isolate it from the rest of the herd for six weeks; this

GOAT HEALTH CHECK

OBSERVATION	SIGNS OF HEALTH	SIGNS OF ILLNESS
Appetite Water usage	Appetite normal Interested in food Drinking normal	Won't eat or drink Too much interest in food Drinking too much water
Attitude Alertness	Bright and alert Inquisitive Normal behavior	Hunched back Moaning or crying No interest in surroundings Staring into space Tail drooping Tremors or shaking Unresponsive
Body condition	Body condition good	Too fat or too thin
Ears	Normal ears	Shaking head Drooping ears Visible parasites or discharge
Eyes	Clear and bright No discharge Able to see	Cloudy or discolored Sunken, squinting, or shut Discharge, tearing Blindness
Feet, hooves, legs, joints, and gait	Stands comfortably Moves easily Puts equal weight on feet	Pain or swelling Limping or lameness Unwilling to stand
Lymph nodes	Normal	Swollen or lumpy
Manure	Normal pellets	Pellets too dry Watery stool or mucus present Feces bloody

is a live vaccine that will infect the entire herd. Some vaccines may also interfere with disease testing. Once a goat is vaccinated against CL, the animal will always give a positive or suspect CL test. CL infection or vaccination has also been reported to cause false positives for Johne's tests.

DEALING WITH SICK GOATS

Nothing strikes terror into the heart of an animal lover more than illness or injury to one of his or her charges. This topic alone fills books. A number of excellent resources are devoted to the medical care of goats. Web pages and goat discussion forums abound.

OBSERVATION	SIGNS OF HEALTH	SIGNS OF ILLNESS
Mucous membranes (eyes and gums)	Pink Moist	Pale Dry Red or off-color
Respiratory	No abnormal sounds Clear or no nasal discharge	Rasping breath Rapid breathing Abnormal cough Green or cloudy nasal discharge
Skin and coat	Skin supple Smooth, silk coat	Skin dry and flaky Coat dull or hair falling out Wounds or lumps
Teeth and mouth	Teeth good Breath normal Mouth and tongue normal	Teeth missing or broken Grinding teeth Breath smells abnormal Scabs or sores Swollen tongue
Udder	Normal shape and texture Normal production Milk white and sweet	Abnormally swollen or hot Sudden drop in production Milk bloody, gassy, or watery Off-tasting milk
Urine	Normal color and amount	Blood or crystals Visible dribbling or discharge Straining

Goats are susceptible to the same types of health problems as any living creature. Being in our care, they deserve proper treatment during illness. Basics include a dry, clean environment; food and water; and safety from harassment—perhaps a separate pen—when they do not feel well. Because a herd always has a pecking order, sick animals may get mistreated by lower-ranking herd mates when they are suddenly vulnerable.

Treatments range from holistic to traditional. I have seen goats left alone and seemingly dying that have recovered to live many more years. I have also attentively nursed goats for weeks using the best care available, yet they still died.

GOAT MEDICINE CHEST

- Alcohol or alcohol preps
- Bandage materials
- CMT test kit (to test for mastitis)
- Castrating supplies (surgical scissors, bander, or Burdizzo)
- Clippers and supplies for clipping
- Collars and leads
- Disbudding iron
- Drench gun or syringe
- Eye puffer or ointment
- Electrolyte replacement powder or fluid (if Gatorade—*not* sugar-free)
- Fecal test kit and microscope
- Feeding tube for kids
- Gloves, leather
- Gloves, surgical, long and short
- Hoof trimmers
- 7% Iodine solution
- Needles—18-, 21-, or 22-gauge × 1"
- Medications suitable for your management style and local recommendations
- Measuring tape or weight tape
- Nolvasan (residuals said to last for two days) or disinfectant
- OB lube (J-Lube powder)
- Peroxide
- Scissors, bandage and surgical
- Splints
- Stanchion and head gate
- Stomach tube for adult goats
- Syringes, 3cc and 12cc
- Thermometer
- Udder infusions
- Vet wrap
- Wound ointment or spray
- Weak kid syringe and stomach tube
- Wormer

In the strictest form, organic practitioners do not use antibiotics or other chemical treatments. Other people put as much medication into the goat as they can afford. I favor a middle ground.

OBSERVATION AND EXAMINATION

Spending time with your goats each day is the best way to keep ahead of health problems. Goats instinctively try to hide an illness as long as they can. Knowing your goat's normal behavior will help you detect problems early on. If something seems off with your goat, it very well could be.

TEMPERATURE

The normal temperature of a goat ranges from 101.5 to 104 degrees Fahrenheit. Caprine temperature varies with the weather as well as by type of goat. Heavy hair coats make a goat warmer in hot weather. The best way to know what is normal in your herd is to check the temperature of several goats so that you can tell how the average animal is running.

A high temperature indicates infection, often bacterial, which means that your goat may need antibiotics. A fever is part of the body's natural defense as it tries to "burn out" the offending agent. Certain medications can help bring the goat's temperature down. If the high temperature stems from an untreated infection, however, the cooler temperature could allow the infection to worsen.

A digital thermometer is the easiest to read and not very expensive. Be sure to get a rectal thermometer. Use a light coating of lube or oil. Holding the goat's tail, insert the thermometer end into the anus. Hold the thermometer in place until it signals that the temperature has been taken. Remove, read, and record the goat's temperature.

Temperatures below normal occur in critically ill animals and chilled newborns. Return the body to its normal temperature with heating pads, heat lamps, or even a hot bath. To warm a chilled kid, wrap the kid in a plastic bag to keep it dry and submerse it to its neck in hot water.

RUMEN MOTILITY

The rumen is a major organ in the goat. Rumen disorders are the root of most critical caprine illnesses. Digestive upsets can become life-threatening without warning. Knowing how to check rumen motility, or movement, is a useful skill. When the rumen is hard and rounded or sounds like a drum when thumped by a few fingers, the goat is bloated.

The normal rumen contracts one to four times per minute. Less frequent motility occurs when the stomach is at rest with very little recent food inside it. Faster motility occurs after a meal. Here's how to measure rumen movement:

1. Put your fingers on the left side of the goat, between the ribs and hips in the soft hollow below the loin.
2. Feel for a hard mass—that is the rumen's contents.
3. Hold your fingers in place until you feel a rolling contraction.
4. Start timing until you feel the next movement.
5. Count the contractions for one minute.

MEDICINE

Goats are a minor species in the farming world and often placed in the generic category of small ruminant, alongside sheep. Due to economics, drugs are rarely tested or labeled specifically for goats. Some medications, most notably wormers, must be given to goats at double the dosage prescribed for other species to be effective. Depending on whom you talk to, a particular drug may be deemed very effective for goats or useless. Along with recommendations from your veterinarian and advice from other goat owners, your observations will help you decide what works best for your herd.

There are several recipes and treatments you can make to benefit your goat's health from supplies in your garden or kitchen cupboard. I have collected these recipes over the years and find them helpful in a pinch.

Of all the possible ways to prevent illness, simply spending time with your goats in one of the best. *Shutterstock*

ELECTROLYTE REPLACEMENT FLUIDS

When a goat refuses to eat, it may become dehydrated and its electrolyte balance could be disturbed. This can also happen when the goat has extreme diarrhea. Dehydration is very dangerous to a goat and should be treated as soon as the condition is recognized. A 100-pound goat needs a gallon of fluids a day. When a goat goes off feed, one of the first possibilities to consider is dehydration.

- 1 gallon warm water
- 2 teaspoons table salt
- 1 teaspoon baking soda
- ½ cup honey, Karo syrup, or molasses (never cane sugar)

Allow the animal to drink free-choice or drench (as described on page 141) conscious animals with small amounts at frequent intervals until interest in food and water returns.

MAGIC SUPPLEMENT

Another homemade potion, Magic is a quick energy supplement. Originally, I made this for does who had just kidded as a pick-me-up or to treat early ketosis. Magic helps with iron replacement and energy. It can be given in any situation where the doe is run down and in need of nutrients.

- 1 part molasses
- 1 part corn oil
- 2 parts Karo syrup
- Dose: 120cc two times a day

WOUND OINTMENT

This ointment can be used in place of many of the wound sprays available from livestock suppliers. It covers the gamut from bacterial protection to ringworm treatment and is very soothing.

- 1 medium container of Vaseline
- 1 large tube of diaper rash ointment
- 1 tube of women's yeast infection medication

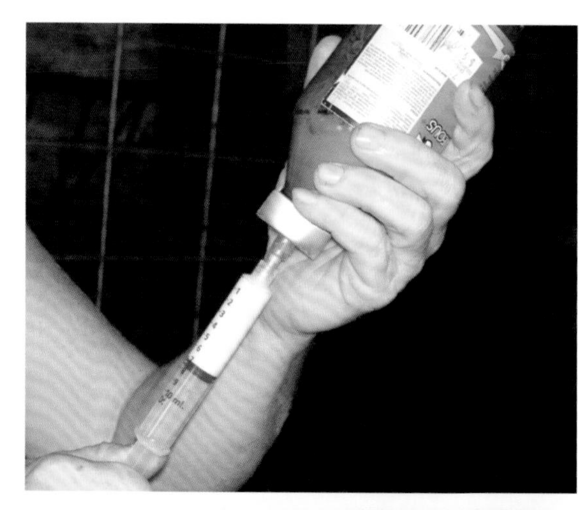

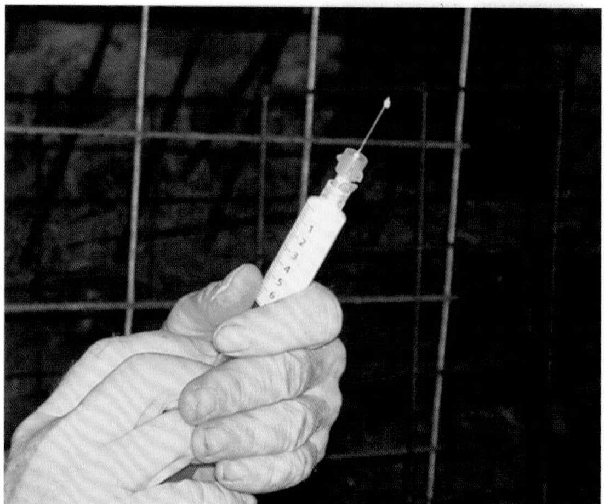

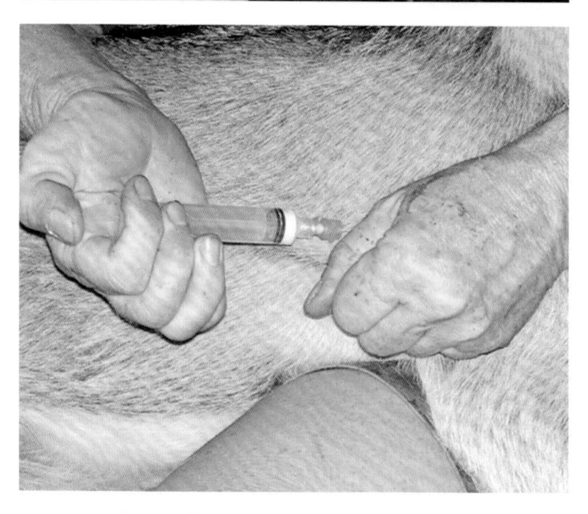

To give medication by injection, you need to fill the syringe until you have the proper dose (top), remove air from the syringe by gently depressing the plunger until a drop of liquid appears on the end of the needle (middle), and inject medicine slowly into a small tent in the skin (bottom).
David Weber, Cutter Farms

ADVERSE REACTIONS TO SHOTS

Sometimes a goat has an adverse reaction to an injection of medication or supplement. Epinephrine can save a goat's life when the reaction is severe enough to cause anaphylactic shock. Epinephrine must be given at the first sign of a reaction, so keep a bottle handy.

REACTION	CONCERN	TREATMENT
Lumps and swelling of injection site	Pain Unsightly Possible to confuse with CLA lumps	Rub injection site gently after giving the shot Ice the swelling Give anti-inflammatory medicine Always inject behind the front leg
Lameness		Give anti-inflammatory medicine
Dragging leg	Paralysis of the nerve Usually associated with penicillin given in the hind leg	Give anti-inflammatory medicine Avoid injecting goats in the hind leg or rump
Rash or raised bumps seen as little spots of raised hair all over goat	Allergic response or mild anaphylaxis	Give epinephrine Give Benadryl
Difficulty breathing Trembling Sudden collapse	Anaphylactic shock	Give epinephrine (an injection under the tongue works the fastest in the case of a severe, immediate reaction)

- 1 tube athlete's foot medication
- ¼ cup Nolvasan or Betadine liquid
- 1 tube triple-antibiotic wound ointment

Warm gently to liquefy, blend, and allow to cool.

INJECTIONS

Many medications and vaccinations are given by injection. Daunting at first, this simple procedure can be learned by any goat keeper. Shots given in the muscle are called intramuscular (IM); shots given under the skin are called subcutaneous (sub Q). An increasing number of injections these days are sub Q. The goat can absorb medication from a sub Q injection without the muscle damage and scarring of IM shots—especially important in meat goats.

Most farmers give shots behind the elbow, where goats do not have lymph nodes. This location prevents injection lumps, which are caused by reaction to the shots, from being mistaken for CL lumps, which occur in the lymphatic system of diseased goats. This is an important consideration for showing goats.

A sub Q injection is given by pulling up a little pinch of skin into a "tent." The needle is inserted into the side of the tent. Take care not to go through the skin on the other side. A gentle pullback on the syringe plunger will tell you whether you are under the skin and can go ahead with the shot. Air in the syringe means you have gone through the skin and must reposition the needle. Blood in the needle means you have gotten into a blood vessel and should readjust.

An IM injection is usually given into the large muscle in the lower hip; the shoulder is a secondary site. Goats should not be given shots in the rump. To be certain you are not in a blood vessel, pull back the plunger and check for blood in the syringe. Incorrect injection into a vein can be hazardous and may even kill the goat, so repositioning the needle is important when blood appears.

ORAL MEDICATIONS

Oral medication may be administered in food or water, as a drench, or as a bolus. A drench is liquid medication that is administered down the goat's throat. A bolus of oral medication is a large pill.

The easiest method, of course, is mixing the medication with food or water. The ailing goats have no choice but to consume it or go without food or water. The problem is that this method is imprecise. Some goats inevitably get too much medication while others do not receive enough. An individual bolus can sometimes be gotten into a goat by offering it wrapped in something tasty. People report success with Fig Newton cookies, dates or figs, or a bit of bread with peanut butter rolled around the pill.

An oral syringe can be used by itself or attached to a stomach tube. *David Weber*

Administering oral medication directly to the goat assures proper dosage, but only if you get the medicine inside the goat! Getting a liquid drench down the goat's throat—as opposed to all over you, the goat, and the floor—can be tricky. And trying to insert a pill into a goat with your hand is a great way to get bitten by its sharp back teeth. Goats are adept at resisting pills. Usually, after dislodging the offending hand, the goat will then contemptuously spit the pill at your feet!

To combat these issues, restrain the goat. Options include holding the goat firmly between your legs, having a helper hold the animal, or using a head gate.

To administer a drench, you will need a commercial drenching gun or a simple needleless syringe. The technique is the same with either implement. To administer a bolus, you need a balling gun. This implement is available from vet suppliers, it comes in several different sizes, the largest of which works better for cattle than goats. The balling gun gets the pill far enough down the animal's throat to force it to swallow. Care must be taken not to accidentally place the pill into the animal's lungs.

STOMACH TUBES

Drenches should only be used in animals that are alert enough to swallow. Weak or unconscious animals need a stomach tube to prevent aspiration of liquid into the lungs. Most often, a stomach tube is used with young kids, especially to give colostrum to weak kids or those that will not take a bottle within the first few hours after delivery. This extremely useful tool consists of a flexible catheter and a large syringe.

For adult goats, livestock supply houses carry large-animal stomach tubes. A large syringe or funnel is used to deliver the fluid. Do not try to use a garden hose to tube an adult goat, it is too large and can kill the goat if it damages the esophagus. For a homemade stomach tube, a 3-foot- to 4-foot-long piece of ¼-inch- to ⅜-inch-diameter plastic tubing rounded off at one end will work. Thread it through a piece of PVC plumbing pipe to put between the goat's teeth to prevent the tube being bitten open.

To administer a drench, the dosing syringe is inserted into the goat's mouth, usually with some objections from the goat. Make certain the goat is swallowing the liquid and it isn't entering her lungs. *David Weber, Cutter Farms*

It is important to be certain you have threaded the tube down into the stomach rather than the lungs. Aspiration pneumonia or drowning can occur when the lungs are suddenly loaded with liquid.

How to Use a Stomach Tube

Here are the supplies you need:
Stomach tube
Marker
Short length of PVC pipe
Funnel

1. Measure the tube against the outside of the goat from nose to stomach to estimate how far it needs to be inserted. Mark the tube.
2. Restrain the goat and hold its head upright. For older goats, put a short length of PVC or metal pipe into the mouth so that the goat won't bite the tubing.
3. Thread the tubing through the pipe (or directly into the mouth) and gently guide it down the esophagus. If something blocks the tube's progress, pull it out and start again.
4. When the tubing is in the stomach, gurgling sounds indicate you are in the right place. Coughing or air sounds may mean you have entered the lungs through the trachea; withdraw the tube and try again.
5. Attach a funnel or the outside part of a 60cc syringe with an irrigation tip.
6. Hold the end of the tubing well above the goat's mouth to ensure good gravity flow.
7. Pour the liquid slowly into the funnel, keeping the tube straight. Gravity will pull the liquid into the stomach.
8. To ensure all the fluid has drained, wait several seconds after the last of the liquid has gone down the animal's throat.
9. Before removing the tube, cover the end with your finger to prevent liquid from leaking and entering the lungs. Remove the tube slowly.
10. If the animal is weak or unconscious, prop it up on its chest. Never lay a goat on its side after tubing or it could aspirate fluid.

Administer only small amounts of liquid at any one time. Some breeders dip the tip of the tube into warm water to soften it before insertion. Another hint: pour a small amount of water into the tube at the end of feeding to rinse the tube before withdrawal. Water, rather than a foreign fluid, is easier for the goat's lungs to handle and less likely to cause pneumonia.

A bolus is a large pill administered with a balling gun (right). The balling gun needs to be gently inserted as far down the goat's throat as possible, then the plunger is depressed, releasing the pill (below, left). Afterward, hold the goat's mouth closed and rub its throat so it will swallow (below, right). *David Weber, Cutter Farms*

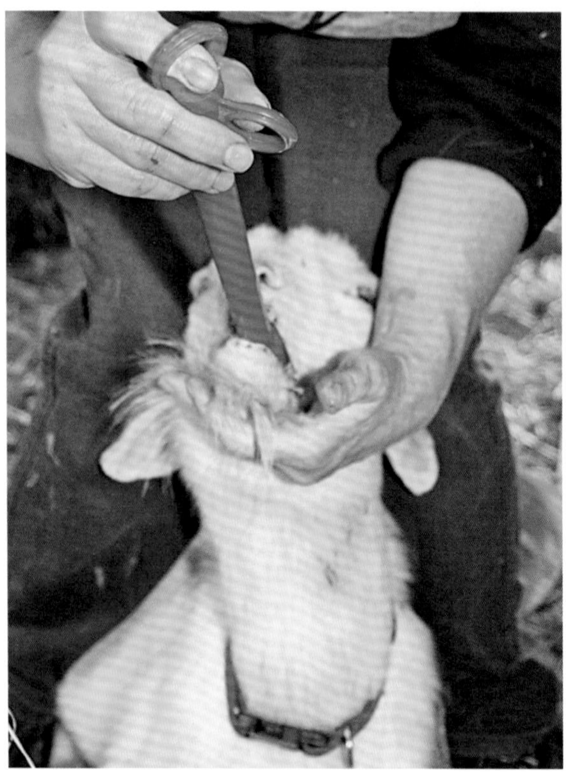

HUMANE SLAUGHTER

Goats quickly lose the will to live when they are in pain. When a goat is too deformed, ill, or injured to survive, the kindest option may be to euthanize. The word euthanasia literally means "good death" in Greek. Discuss euthanasia options with your veterinarian and create a plan as part of your emergency medical kit. A vet can be called to the farm to give a lethal injection. When a vet is not available for this service, the herd owner may need to perform the task.

As stressful and unpleasant as it may be, preventing unnecessary suffering of the animal should be paramount in any situation with sick goats. When performing euthanasia, follow these guidelines: Goat skulls are very strong in front, so a gunshot should be taken either from the back of the skull between the ears, with the muzzle of the gun pointing downward through the mouth and between the teeth, or just behind the ear, aiming toward the opposite eye.

- In the absence of a gun, a traditional ritual slaughter technique may be used by cutting the throat straight across both jugular veins with a very sharp knife. The rapid blood loss is effective but messy and disturbing.
- Suffocation or drowning are not recommended.
- Barbiturates and anesthetics are available only to veterinarians. It is unacceptable to use over-the-counter products as a substitute for veterinary chemicals.
- Confirm that the animal is dead by checking whether the heartbeat, respiration, and corneal reflex are all absent.

If you intend to perform euthanasia yourself, know your local laws in order to avoid prosecution under animal-cruelty statutes. What is standard practice in one part of the country may be considered cruelty in another, so do not assume.

A number of years ago, a prominent dairy-goat owner was charged with felony mistreatment of animals for putting down terminal kid goats using a blow to the head. While those charges were later dropped, the adverse publicity and $5,000 in legal fees were extremely damaging.

FINDING A VETERINARIAN

One of the goat owner's biggest challenges can be finding a competent veterinarian. Goats are not covered to a large degree in veterinary school. To find a good caprine vet, you'll need to talk to local goat owners and possibly interview a number of vets.

I consider myself lucky. I live in an area with several highly knowledgeable and skilled practitioners. Still, I called a few vets before settling on my current one. One veterinarian, who shall remain nameless, gave me a lecture about why he won't treat goats. He really wanted me to know how much of a pain he found hobby-farm goat owners!

To gauge the vet's overall skill, watch how the doctor handles the animal. Is care taken to be gentle? Willing to discuss treatment options with you? Knowledgeable about basic goat vaccinations and diseases? Willing to admit lack of knowledge and look up the information when unsure?

If you see something that bothers you, question the vet. Do not hesitate to request that the vet either change the treatment practice or leave your farm. The response will enlighten you as to whether you want to use the vet again.

Sometimes an incident or a mishap will leave no doubt. One farmer quit using a vet after the doctor threw the contaminated contents of an abscess directly onto the barn floor after removing it from a goat. And I suspect that a vet infected my first two goats with CAE during a routine blood test. He used a needle that had been used previously on other goats at the farm where I took them to be tested.

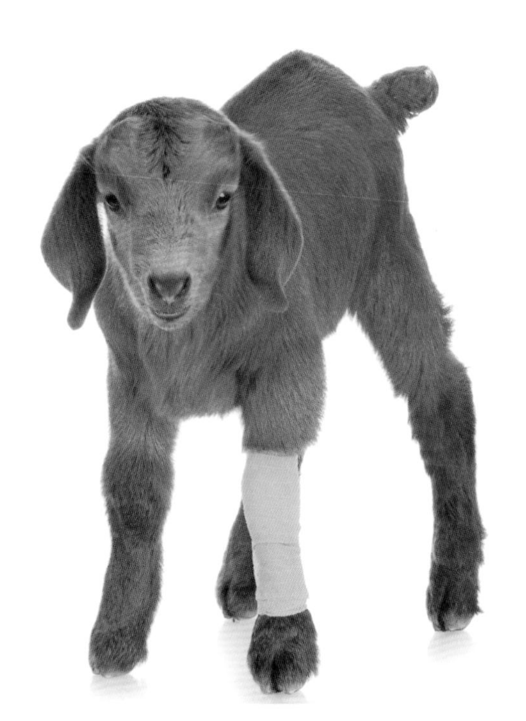

Neutering, spaying, disbudding, and other preventative techniques may seem cruel to a kid, but they recover quickly and are often better for it.

In another alarming case, a vet performing a difficult delivery finally managed to extricate the kid. The vet tossed the kid against the wall of the barn, stating, "That one is dead!" The horrified owner retrieved the kid, swung it gently by the hind legs, and revived it. She never used that vet again!

On the other hand, do not judge a vet by things that go wrong in the normal course of farm life. One of our goats died in elective surgery to remove an abscess located close to the jugular vein. A bleed occurred, and two vets worked very hard to save her, but she died anyway. They did their best in a bad situation.

Your veterinarian is your resource for prescription medications, off-label drug use, and myriad medical skills that the basic goat owner can't begin to learn. Labels can also quickly become out of date, since medical advances happen so quickly. A good veterinarian has the resources to find current information.

Finding the right veterinarian to care for your herd is vital to long-term success as a goat farmer or breeder. Take care to ask questions, observe his or her actions with your goats, and that proper and sterile medical techniques are practiced.

Be aware that it is against the law to use off-label drugs without the advice of a veterinarian. The Animal Medicinal Drug Use Clarification Act of 1994 allows vets to use certain drugs "in a manner that is not in accordance with the approved label directions" if the owner and vet have a client-patient relationship.

CHRONIC CONDITIONS

Chronic illnesses may start with an acute condition or may lie dormant and never show in an individual goat. Some of the most troublesome current diseases are CAE, CL, Johne's, and chronic wasting diseases, such as scrapie.

Conditions including brucellosis, tuberculosis, and bluetongue may be tested for depending on where you live and where your goat is going if you're preparing to move it. Some buyers request testing prior to buying a goat. Diseases such as tuberculosis in the United States and hoof-and-mouth in Great Britain must be reported to health agencies. Often, animals with reported infectious diseases are subject to quarantine and disposal.

Some genetics are known to protect goats from certain diseases. A test is now available to test for scrapie resistance, following the success of a genetic test used in sheep. A genetic disorder currently being studied in Nubian goats is mucopolysaccharidosis IIID, or G-6-Sulfase deficiency (G6S). As this disorder becomes better understood, Nubian breeders and buyers are testing for G6S with increased frequency.

CLOSED HERD

The strictest form of disease management involves testing and culling any positive animals via slaughter. Never sell a goat that has tested positive for a disease unless you fully disclose the animal's condition to the buyer. The goal of this management style is disease eradication.

To ensure no diseases are brought to the farm, a strict closed herd does not buy animals, show animals, or allow other goats onto the premises for any reason. Visitors to closed herds may be asked to wear clothing that has not been in other livestock facilities or to use wash stations for footwear. Some herds modify these restrictions to allow showing and the purchase or breeding of "test negative" animals. Most closed herds test all animals on a regular schedule to catch new infections and any false negatives from previous testing.

Buying an animal from a closed herd maximizes your chances of getting healthy stock. Not surprisingly, these goats are normally costlier. In some areas, prices haven't necessarily increased at the same rate as expenses. In my opinion, herds should ask for and receive better payment for extra disease prevention methods. The costs of testing, quarantine, culling, and so on add up quickly. The price of such goats should reflect the expense of veterinarian visits, blood testing, and breeding fees.

ISOLATION

Even serious diseases such as CAE and CL do not have to be a death sentence for the goat. Many goats have lived long, productive lives after receiving positive test results. Isolation achieves the ultimate goal of eradicating the disease in the herd while recognizing the difficulty of putting down asymptomatic, well-loved, or valuable animals.

Some owners keep two separate herds: one with positives and one with negatives. All "test positive" animals are handled last, with different equipment, to prevent inadvertent spread of the disease. To minimize the chance of spreading disease from the positive herd to the negative one, hygiene becomes paramount. Caretakers need to wash well and use different clothing and boots when moving between groups. All kids, or sometimes just those from positive does, are bottle-reared on pasteurized milk.

COMINGLING

Some goat owners choose to let goats of unknown disease status or "test positive" and "test negative" goats comingle in the same herd. The cost of disease testing is eliminated by accepting that animals aren't checked for disease status unless a problem arises. If you have a small herd or less time or space to devote to your herd, this may be the best option. Of course, goats that have visible symptoms or open abscesses should be removed from the herd by culling or kept isolated until open wounds are healed to prevent further spread of the disease.

Often, the goal of this type of management is to raise disease-resistant, hardy animals with strong immune systems. The number of potential buyers may be limited to these herds if buyers expect to purchase animals without underlying conditions. While I no longer test my herd, I offer buyers the option of paying for testing at their expense.

COMMON CAPRINE DISEASES AND AILMENTS

ABORTIONS

Toxoplasmosis, chlamydiosis, vibriosis, and brucellosis are infectious diseases that can cause spontaneous abortions. Pesticides, wormers, and other medication may also cause abortions.

One or two spontaneous abortions a year in a large herd is not usually cause for panic. An "abortion storm" is a situation when several, or sometimes most, of the does in a herd spontaneously abort in the same season. This is heartbreaking.

Toxoplasmosis is spread by young cats. Older animals develop resistance in the form of antibodies to the organism. Once a person or animal develops antibodies, they are immune. A few years after I got my goats, we had an abortion storm in which multiple goats lost twins, triplets, and even quads. It was difficult to go to the barn in the morning and find them. After I took samples to the University of Minnesota Veterinary Diagnostic Laboratory, the problem was diagnosed as toxoplasmosis. We never had another outbreak. Quite a few years later when I became pregnant with my daughter, the doctor tested my blood. I had a high titer of antibodies, which indicated that at some time, I also fought off and won against toxoplasmosis.

Care should be taken when working with delivery fluids, fetuses, or afterbirth. Toxoplasmosis and chlamydiosis are transmissible to humans. Because abortion has so many possible causes, a veterinarian should be consulted. Save the dead kids and

Toxoplasmosis is a disease that is transmissible to goats, other animals, and even humans. It is carried by barn cats through their waste. Pregnant does who contract "toxo" may abort or have deformed kids. Keeping cats out of feeders, neutering them, and keeping only mature felines on the farm will minimize this risk. *Jen Brown*

placenta in a cool location until you can reach a veterinarian. Often, the best way to diagnose the cause of the infection is an examination of the placenta. Your vet will then be able to tell you the next steps to take.

BLOAT

Stomach issues are some of the most common worries when caring for goats. Bloat is one of those problems that is better to prevent than to have to treat. The goat's rumen processes food by fermentation. When excess gas builds up too rapidly or something blocks the gas, it cannot be expelled. This build-up of gas fills the stomach and the goat will bloat. A digestive problem like this must be resolved quickly or the goat can die.

Indigestion is caused by eating too much grain or fresh greens and can occur before or following bloat. The goat acts colicky, stamps its hind feet, acts depressed, grunts, and bites at its sides. An affected animal may even lie down, generally acting sick. This behavior is usually followed by diarrhea.

In any type of bloat, symptoms include a distended abdomen, piteous crying and moaning, grinding of teeth, depression, and obvious discomfort. I've seen goats kick up at their stomach with their feet or lie on the ground with feet held straight out and head thrown back. The goat's left flank bulges with gas. If thumped, the stomach sounds hollow. Kids may be almost round and will slosh when jiggled. Goats may be found down and almost dead from respiratory or circulatory failure.

Reviewing goat literature or exploring the internet will yield a variety of old and new remedies. Some of them work and some don't. It is important to know the kinds of bloat, have treatment on hand, and act quickly or contact your veterinarian for immediate treatment. With any bloat, remove the animal from the source of feed causing the problem.

Frothy bloat is more dangerous than dry bloat. It is caused by overeating lush, green plants or moist feed. Froth fills the rumen with tiny bubbles. These bubbles foam together and become impossible for the goat to release in a belch. An affected animal can die quickly from the excessive pressure on the diaphragm.

Breaking up this foam is the first step to solving frothy bloat. Some treatments for this include the following:

- Di-gel or simethicone products to eliminate the froth and allow the goat to belch
- Therabloat, an oral product for cattle used to break up the bubbles, given orally—3cc well-mixed with 30cc water
- Milk of magnesia, which will lower the pH in the rumen and stimulate the gut—especially useful when grain overload is suspected
- A handful of sodium bicarbonate, which adjusts rumen pH and helps decrease gas

My go-to treatment for bloat has always been to drench the goat with 1 to 2 cups of vegetable or mineral oil. This method is used successfully by countless goat owners and is widely recommended. But the oil isn't as effective in breaking up gas as the other treatments listed here. I have relegated oil back to my kitchen cupboard and recommend other treatments instead.

Another old home remedy uses powdered laundry detergent (usually Tide), dish soap, or Jet-Dry for curing bloat. These suggestions rely on the surfactants in the products to breakup bloat. Because soaps are not made to be ingested and may contain unknown chemicals, I prefer to avoid this treatment except as a last resort.

Once you have given the goat any of these treatments, it helps to break up the gas by gently rolling the animal back and forth on the ground or jiggling a kid up and down on your lap. Walk the goat to encourage rumen activity. If the goat is resting, place its front feet uphill so the gas can move upwards and come out in a belch. Feeding dry, coarse hay will stimulate the rumen.

Oral penicillin is very effective against clostridial infections. As a result, some people give oral penicillin to treat bloat as well as following a gas-reducing protocol. The antibiotic will also kill off beneficial bacteria, so probiotics or yogurt can help restore gut function when the bloat has subsided.

Dry or Free Gas Bloat

Dry bloat is usually brought on by binge eating, oftentimes grain. Closely monitor any goat that has pigged out in the family garden or gotten into the feed room. Sometimes it occurs as a byproduct of some underlying illness. Gas forms in large bubbles or pockets, becoming trapped in the upper part of the goat's rumen. As gas continues to form, if the animal can't belch—it bloats.

Do not give a goat with dry bloat any water. If it has eaten too much grain, water will cause the grain to expand and increase the fermentation rate. Withhold water for about twelve hours. Give the goat roughage to help stimulate the rumen. Give the goat oral bloat medications such as Gas-X.

If the bloat is very severe, a stomach tube can release the gas effectively. This only works in cases of dry bloat. Frothy bloat cannot be relieved with a stomach tube, which only works on larger pockets of gas.

Choke

During choke, something lodges in the goat's throat or esophagus. The blockage prevents belching and a backup of gas occurs. It is especially serious when a goat swallows something large enough to block the whole esophagus. If the obstruction can be seen, use leather gloves to protect your hand and try to gently pull the item free. Sometimes, a stomach tube passed down the throat will push the obstruction free and into the stomach. Otherwise, veterinary assistance is critical.

BLOAT PREVENTION

- Keep feed in tightly sealed, goat-proof containers or rooms.
- Feed dry hay before letting goats eat high-moisture pasture such as clover or legumes.
- Limit access to fresh, new pasture.
- Offer baking soda free-choice in a mineral feeder.
- Be careful not to give excess treats the goats aren't used to, such as green corn stalks or vegetables, especially brassicas like cabbage or broccoli.
- Freshly put up alfalfa or clover hay bales should not be fed warm.
- Feed kids smaller, more frequent meals.
- Free-feed cold milk, which encourages kids to eat less at a time.
- Mix any powdered milk replacers thoroughly and use non-soy-based product.

EMERGENCY BLOAT PUNCTURE

Drastic measures are required if your goat does not respond to normal treatment or is found in extreme distress—difficulty breathing or completely down. I strongly recommend a veterinarian in this situation. However, distance or lack of a vet may leave you with no choice but to release the gas yourself. The difference between saving the goat or losing it can be only minutes. Thankfully, I have never had to perform such an operation.

A 1½-inch 16-gauge needle can be used to make a puncture into the rumen through the highest part of the inflated flank. Using a knife or scalpel in an extreme emergency will also work. This method creates a wound that will need surgical repair. Either of these operations risks leaking fluids from the rumen into the peritoneum. That leakage is full of organisms that can cause a severe infection.

Insert your 16-gauge needle on the left side of the goat midway between the last rib and the hip. This puncture should be at the highest point of the bloat. Push the needle as far into the bulge as possible and hold it in place while the gas escapes.

BLUETONGUE

Bluetongue viruses in the United States have long prompted officials to block the export of American goats (as well as cattle and sheep) to many world markets, such as Australia, New Zealand, and Europe.

Transmitted by insects, bluetongue in goats is mild. Signs include inflammation and bleeding of the mucus membranes of the mouth, nose, and tongue, plus soreness and swelling of the feet. The condition is called bluetongue because of the color of the tongue or mucus membranes.

BROKEN BONES, SPRAINS, AND STRAINS

Goats have accidents—breaks or sprains are not uncommon. Sometimes, goats twist a foot or a leg in a hole in the ground or catch a limb in a fence panel or feeder. Jumping off objects or being rammed by another goat can also cause injury. Examine any goat that is limping or favoring a leg.

A strain or sprain may be treated with liniment rubs and anti-inflammatory medications, such as aspirin, willowbark, phenylbutazone, or Banamine. If you have experience, you may choose to wrap the limb with splints and Vet Wrap by yourself.

Unlike horses, goats can recover nicely from broken bones. Your veterinarian will x-ray the goat and then recommend treatment.

BRUCELLOSIS

Brucellosis is an infectious disease caused by *Brucella* bacteria. Sheep, goats, cattle, pigs, and dogs—as well as wild animals like deer and elk—are vulnerable. Humans can become infected by drinking infected milk, eating undercooked meat, or contact with infected animals. Known as Ungulate or Malta fever, symptoms of infection in humans include flulike fever, sweating, headaches, back pains, and physical weakness. Severe cases may become chronic. In goats, brucellosis causes abortions, weak kids, and production losses.

The US Animal Health Association once believed that the country was free from *Brucella melitensis*. However, a cow in Texas tested positive for brucellosis in 1999 and resulting testing of neighboring farms discovered the disease in a herd of meat goats. The animals were destroyed. Since brucellosis is still found in Mexico, herds along the US-Mexican border are at risk, although the risk is small.

Brucellosis is a commonly included in routine testing for goats being transported or used in commercial dairies. Some shipping or commercial dairy regulations require testing for brucellosis because of the zoonotic nature of this disease.

CAPRINE ARTHRITIS ENCEPHALITIS

Caprine arthritis encephalitis, or CAE, is credited with causing significant losses to dairies and breeders. Estimated milk production decline from CAE are as high as 25 percent. Productive animals are culled or lost through pneumonia, arthritis, and increased susceptibility to other diseases and parasites. Australia has a voluntary program for the elimination of CAE in its herds. Norway and Switzerland are reported to be free of the disease.

Viral leukoencephalomyelitis, or VLE, was a nervous disease first diagnosed in goats in 1974. By the early 1980s, arthritis was linked to the disease. The name was then changed to caprine arthritis encephalitis. Additional conditions such as hard udders or pneumonia have now been proven to be caused by the CAE virus as well.

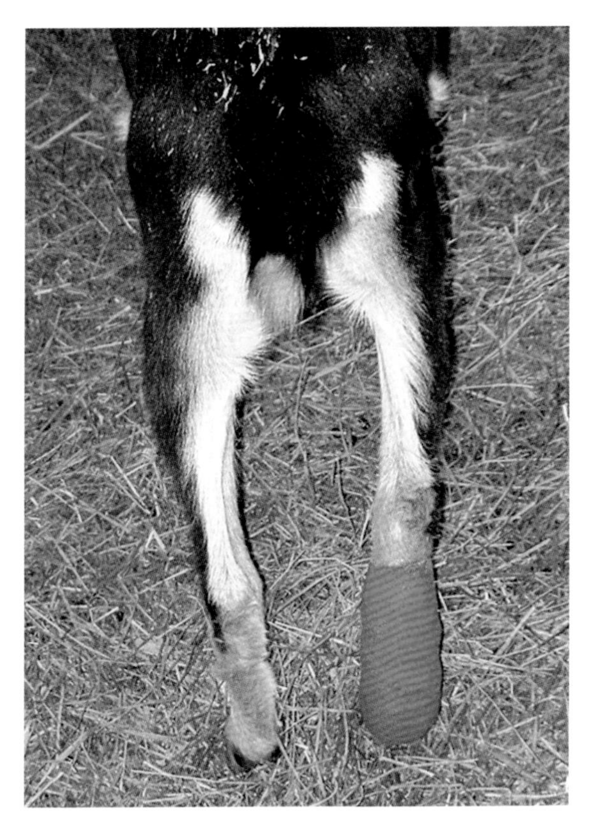

Broken limbs can be splinted and wrapped in Vet Wrap, allowing the goat to heal. *Carol Amundson*

A lentivirus like HIV in humans and feline leukemia in cats, CAE lives in the WBCs and passes through body fluids. CAE is only transmissible to goats. It is most common in dairy goats because of the practice of feeding pooled milk.

People have used infected milk in the mistaken view that it can treat AIDS (HIV). While there is cross-reactivity between CAEV and HIV in immunological testing, there is no scientific validity in the belief that CAE-positive goat milk is either a treatment or preventative for AIDS.

Goats infected with CAE may never show signs of illness. Rarely, kids between two and six months of age develop a fatal encephalitic form of the disease. CAE's primary visible signs appear in older goats as swollen knees or lameness leading progressively to a crippling arthritis. Lowered immune capabilities, interstitial pneumonia, congested udders, and lowered milk production are also linked to this virus.

One common sign of CAE that is frequently mistaken for other conditions is weight loss and poor hair coat. This deterioration may be caused by the inability of the goat to compete for food. This might be a reason that smaller, intensively handled herds often see fewer signs of CAE. Goats under less stress with little competition are reported to have slower observable onset of disease.

CAE prevention is a management practice involving the removal of newborn kids from the dams and raising them separately from the herd. Birthing fluids and milk are the most common routes for spreading CAE. Feeding kids heat-treated colostrum and pasteurized milk or milk replacer prevents them from becoming infected through their mother's milk.

Long-term contact between healthy goats and latent carriers may spread CAE. As research has progressed, the virus has been found in other body fluids such as saliva and semen. Some strictly clean herds eliminate all contact between negative goats and animals that have positive test results or unknown disease status.

Using the same needle to inject multiple goats can spread the virus. The same may be true if bloodsucking insects, such as ticks or lice, move from goat to goat.

During breeding, the risk is greater that an infected doe will pass the infection to the buck rather than the other way around. Either way, there is a risk, so some breeders perform outside breedings only on the condition that the buck and the doe have both been tested for the disease.

Retroviruses live only a very short time outside of living cells. CAE is killed in the environment by drying or by contact with bleach or other antiviral agent. There is no cure or vaccination against CAE. Some comfort can be given to affected goats by treating the symptoms.

CAE is one of three main disease tests often requested by people buying goats—the others are Caseous lymphadenitis and Johne's. Testing should be done only on goats that are at least six months old. Tests done on younger goats or within one month of kidding are not considered accurate.

CASEOUS LYMPHADENITIS

Abscesses, lumps, and bumps can be signs of caseous lymphadenitis (CL), a disease of the lymph nodes. CL results in tremendous cost to commercial herds from the loss of revenue from condemned carcasses, the inability to show goats that have visible lumps, and the loss of sales to owners who want CL-free herds.

The organism that causes this disease, *Corynebacterium pseudotuberculosis*, is highly resistant. No real treatment exists. While infection is most frequently seen in goats or sheep, infections also occur in cattle, horses, pigs, wild ruminants, poultry and even people.

CL appears in goats mostly as lumps or swollen glands in and around the lymph nodes. Internal abscesses are more common in sheep than in goats and lodge in the lymph system, lungs, or other organs. Pus from broken abscesses infects the environment. When there are abscesses in lungs, contamination occurs through nasal secretions or when the animal coughs up infectious phlegm. Other goats can pick up CL from rubbing against common fences, feeders, and other contaminated surfaces. In a show setting, pens that have previously held animals with active CL could infect the goats placed in that pen.

An abscess progresses from a small lump to a ripe capsule that has little or no hair cover. The incubation time for CL is from one to three months for the abscess to completely encapsulate. These ripened abscesses can be cleaned out and flushed with diluted Nolvasan or other disinfectants. A veterinarian can perform surgery to remove all or most of an abscess. Removal may only be a temporary

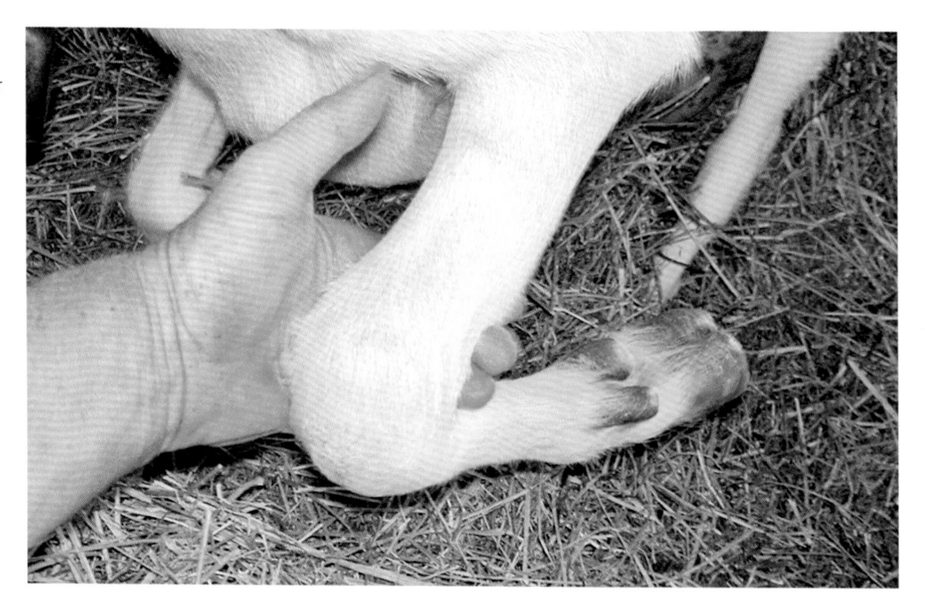

Arthritis can be crippling, especially when caused by CAE. This goat can no longer extend her leg. Often, the first sign of CAE is swollen knees. *Jen Brown*

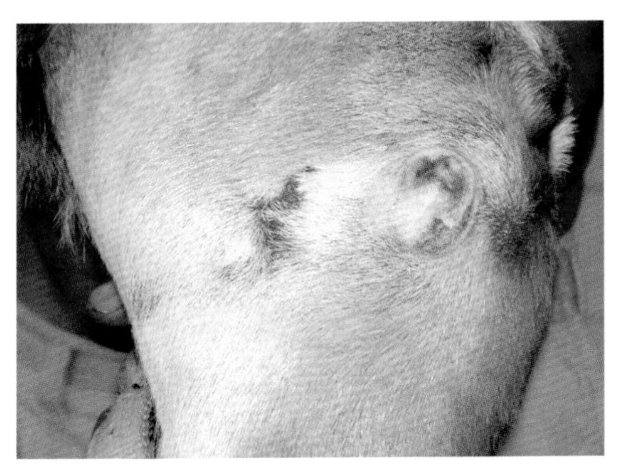

The lump, scars, and healing wound on this goat's throat are likely to be CL, but only a culture can confirm the diagnosis. *Jen Brown*

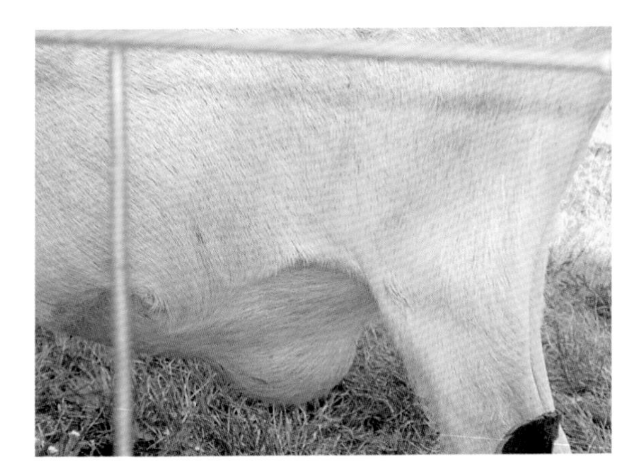

The lump on the doe's abdomen is from an abdominal hernia and is not infectious. *Jen Brown*

Lumps on the neck are a disqualification at most goat shows, as they can indicate CL. Be sure not to administer vaccines in the neck, so the vaccination lumps don't become confused with CL lumps. *Jen Brown*

fix if the abscess was not fully encapsulated before surgery or the CL organism is already circulating in the lymph system. An old procedure using formaldehyde to destroy the abscess is no longer recommended.

Not every abscess indicates CL, but only a culture can tell for certain. Your veterinarian or a private testing lab will use a sample of the pus to grow the organism for positive identification. In some closed herd protocols, it is recommended that all abscesses are cultured.

Goat keepers who have chronic CL problems in their herd may decide to vaccinate. There are commercial generic CL vaccines available. In addition, because there are a variety of strains of *Corynebacterium* causing CL, private laboratories can create an autogenous vaccine. In either case, there have been reports of goats reacting badly to the vaccine due to either pain or dropped milk production. Seek advice from your vet or other local breeders before starting a CL vaccination program.

Breeders need to remember that vaccinated goats will likely test positive for CL because their body has developed the CL antibodies detected by testing. CL infection or vaccination has also been known to trigger false positive Johne's test results as well.

DIARRHEA OR SCOURS

Let's talk poop! I can't stress enough: watch your goats and note changes in behavior, physical condition, and what goes into or comes out of their bodies. Normal goat droppings are round, firm pellets referred to as "nanny berries." The softer or looser the stool, the greater the chance the goat is sick. Minor dietary changes can cause log-style droppings, a pudding consistency, or minor diarrhea.

CAUSES OF DIARRHEA IN GOATS

Bacterial	*E. coli*
	Salmonella sp.
	Clostridium perfringens
Parasites	Stomach worms
Protozoa	*Cryptosporidium* coccidia (*Eimera* sp.)
	Giardia sp.
Diet	Dirty water buckets or those fouled with stool
	Feed changes or too much feed or indigestion
	Poor-quality colostrum or not enough within the first 24 hours Poor-quality or improperly mixed milk replacer
	Rich, wet pasture—especially sudden exposure
	Roughage—too little or not enough dry matter
	Toxins or allergies
Management	Overstocking/overcrowding
(Poor environment)	Poor sanitation
Stress	Changes or extremes in weather
	Changes in routine
	Traveling or shipping
	Weaning
Viral	Rotavirus
	Coronavirus

Some diarrhea is watery and squirts from the sick goat with every cough or movement. This is called scours. It means your animal is very sick and can become dehydrated and possibly die very quickly. To determine the treatment, first determine the cause:

- Review any changes to your goat's diet or feeding environment. Have you changed the type or amount of feed? Has the animal been grazing in a new or different pasture that has a different or more abundant vegetation? Taking your ailing goat off rich grain and switching to just hay will help.
- Consider the age of the animal. Is it a newborn or just a few weeks old? Scours is a common caprine problem during the first few days of life. It may be caused by chilling, erratic feeding, dirty bedding, dirty milk bottles, or overeating. Spectam Scour-Halt (a pig medication) does a good job on very young kids that have diarrhea. Warning: do not use this medication on adult goats, as it works so well that it will destroy the working rumen's natural organisms and shut down the goat's digestive system. Scours in a

three- to four-week-old kid is likely due to coccidiosis.

- Consider the color of the diarrhea. Is it green or black? Green diarrhea may indicate an eating change or plant poisoning; black, tarry stool is a sign of bleeding in the digestive tract—possibly due to coccidia or stomach parasites. Yellow or light brown diarrhea tends to be from bacterial causes.
- Check for internal parasites. Worms can cause scours. Check the stool for the presence of parasites. This can be done by your veterinarian, or you can do it yourself if you have a microscope and reference materials. Proper deworming should take place if parasites are discovered.
- Enterotoxemia type C, also known as "bloody scours," is caused by the clostridium bacteria and usually seen in the first few weeks of life. The disease causes bloody infection in the small intestine.
- Enterotoxemia type D, also known as "pulpy kidney disease," is another bacterial disease, caused by a different strain of clostridium, that affects kids over four weeks old. It often affects the fastest growing, biggest kids.
- Both illnesses can start due to sudden feed changes. Affected kids are usually eating a large amount of grain or nursing on does that produce lots of milk. Adults can also become sick with enterotoxemia after sudden feed or weather changes. Clostridial organisms are common in the environment and in the guts of ruminants. When feed change disrupts the balance of gut flora, the wrong organisms may take over, creating toxins that can cause indigestion or even death. When a herd becomes infected, losses may be high.

Treatment of Scours

Regardless of the underlying disease causing the severe diarrhea, the fluids and electrolytes expelled can cause an animal to become dehydrated and very ill in short order. Any dehydration needs to be treated along with the illness that is causing it. Electrolyte solutions will help prevent dehydration. If the goat will drink, offer these fluids or water frequently, especially if the diarrhea is severe. If you're using a product like Gatorade, do not give diet products. Xylitol and artificial sweeteners can be toxic to animals.

Normal nanny berries are round, firm pellets. Pudding-like poop is abnormal and a sign that your goat is likely sick. *Shutterstock*

ENTROPION

Sometimes a newborn goat's eyelid turns inward, causing the eyelashes to rub against the eye and irritate the cornea. The kid's eye waters, crusts over, or swells shut. When caught early, entropion is easily repaired.

Injecting a small amount of penicillin directly into the eyelid causes the lid to swell and pulls the eyelashes away from the eye. In a few days, the swelling goes down and the eye becomes normal. You can also use a small clip or stitch the lid back after rolling the lashes outward.

Entropion can be caused by too much ultraviolet light exposure or heat lamp exposure, as well as genetic disposition. Because it can be genetic, some people consider entropion to be a culling consideration. I never have worried about this because it occurred infrequently. Good records of kids treated each season will allow breeders to decide if this is true in their herd.

FLOPPY KID SYNDROME

Floppy kid syndrome (FKS) is another problem caused by clostridial organisms.

FKS is usually seen in kids from three to fourteen days old or, rarely, up to a month of age. It is frequently seen in late spring kids as the weather warms up. The organism is found in soil. Symptoms include weakness, wobbly walk, and poor muscle tone—hence the term *floppy*. When gently shaken, the kid's stomach will slosh. It is seen in dam-raised and bottle-fed kids. Often the first sign is a kid crossing its back legs as it walks or runs and then stumbling. The kid may have a dirty mouth from eating soil. Fever or respiratory distress may occur, but not always. No diarrhea is seen.

It is critical when finding a kid down to decide if it is due to FKS or being weak. Kids that are newborn up to seven days old can get separated from mom or may not be nursing often enough. These weak kids may be mistaken for having FKS when they are not getting enough nutrition. In the case of multiples, mom sometimes favors one or two kids while ignoring one. Be sure to check the doe's udder to see if the limp kid seems to have eaten. Cool, limp kids should be warmed up to a body temperature greater than 100 degrees Fahrenheit and tube fed.

When malnutrition has been ruled out, ½ to 1 teaspoon baking soda mixed with a little water is helpful if the kid's stomach is full and nonfunctioning. This may be the only treatment needed if the illness is caught immediately.

Treat kids that scour and are still active for a full three days with oral doses of penicillin, fortified vitamins, and thiamin. Kids that are completely floppy may need to be tube-fed and given an injection rather than oral medication.

Do not tube milk into the stomach of flat kids, but instead use electrolyte and dextrose solutions until the kid has been up and moving for at least eight hours. Putting milk into the stomach of a kid whose temperature isn't at least 100 degrees Fahrenheit will kill the kid.

FOOT ROT

Foot rot and foot scald are hoof infections seen in warm, wet weather. These problems are more common in sheep than in goats. Affected animals may start out with a mild limp. Left untreated, the disease spreads to all four feet and makes the goat unwilling to walk or eat. Muddy confinement areas, crowded conditions, and rainy weather facilitate this disease.

Affected animals should be treated by trimming the hoof and using a foot bath of copper, zinc, or iodine. I have treated individual goats with mild foot rot by pouring a little Kopertox over the affected area several times a day as well as trimming as necessary. For prevention or treatment of larger groups of goats, a footbath tray can be placed at one end of a run on the way into a pen. Forcing the goats through this bath will treat the entire group.

GLUCOSAMINE-6 SULFATASE DEFICIENCY

Glucosamine-6-sulfatase (G6S) deficiency causes a recessive genetic disorder called Caprine Mucopolysaccharideosis-IIID. It is prevalent in Nubians or Nubian-cross goats. Transmission is through a nonsex-linked genetic mutation. In 1987, a thousand goats tested in Michigan showed nearly 25 percent of the goats carried the abnormal gene. Since that time, various studies continue to show a prevalence >23 percent among Nubian goats tested.

If a goat parent carries the trait, 50 percent of offspring are carriers. Carrier animals are healthy. When two carrier animals are bred, there is a 25 percent chance that kids will inherit G6S deficient genes from both parents and have the disorder. G6S-deficient goats lack an enzyme needed for growth and maintaining connective tissues. Animals may die young, or they may live up to four years in weakened condition—allowing them to breed and pass along the genetic defect.

Nubian breeders can avoid the disorder by knowing the genetic status of at least one parent and breeding appropriately. Testing is now readily available through veterinarians or testing laboratories. The American Dairy Goat Association has information for members to send samples for testing to an approved lab. ADGA now maintains a list of tested animals and their status with permission of owners.

Genetic testing for goats is normally performed using blood, semen, or hair. Hairs are collected using a needle-nosed pliers to grasp a hair close to the skin and pull the entire hair including the root. Twenty to thirty hairs are usually enough for the test. Follow the lab's instructions for shipping and handling.

HOOF-AND-MOUTH DISEASE

An extremely contagious viral disease seen in cattle, pigs, goats, and other livestock is hoof-and-mouth disease, also known as foot-and-mouth disease (FMD). Humans *can* catch it in very rare cases; however, the risk by humans, as well as cats and dogs, is in spreading the disease through contact. The virus will spread on clothing, equipment, fur, and so on. FMD will survive in human nasal passages for as much as twenty-eight hours and can infect animals contacted during that time. Affected livestock have blisters and sores around the lips, the tongue, the teats, or the coronary band of the hoof. Goats become lame and sometimes drool. They also have difficulty eating when the sores in the mouth are numerous. Death may occur.

There have been no hoof-and-mouth cases in the United States since 1929 and none in Canada or Mexico since 1954. Currently, animal import bans from affected parts of the world keep this disease out of the United States. However, the extremely virulent nature of hoof-and-mouth was seen by the destruction of many cattle in affected parts of England in recent years. Vigilance is important.

JOHNE'S DISEASE

Johne's disease is a fatal chronic wasting disease of livestock similar to tuberculosis and leprosy in humans. Johne's is also known as paratuberculosis, from the organism that causes it—*Mycobacterium avium* subspecies *paratuberculosis*, or MAP. The bacterium lodges in the intestines of the animal, causing thickening of the intestinal mucosa and blocking the digestion of nutrients.

Affected goats are not always symptomatic, making the disease hard to diagnose. Johne's is spread through the organism shedding in the stool of often asymptomatic infected animals. Healthy animals consume MAP when they mouth hay, bedding, or other items contaminated by feces. The bacterium is also present in untreated milk or colostrum from infected does.

Cattle and sheep, as well as other ruminants, can also be infected by Johne's. As a result, goats purchased from sale barns, where large numbers of animals mix and congregate, are at increased risk. Johne's will remain in infected soil for several years. Older goats have less risk of contracting Johne's than kids. Very young kids in the first months of life are most commonly infected. These kids usually don't show signs of infection for years. Immune-compromised goats with other disorders such as a heavy load of parasites are also at greater risk.

Johne's is a devastating infection to discover in a goat. Often, once the diagnosis has been made in one animal, it will have infected other goats in the herd. Due to the long incubation period, lack of testing, and possibility of confusing MAP with other illnesses, no one knows how widespread it is in the goat world. However, it has been found in all types of herds, from pets to dairy to meat and fiber.

Goats showing active Johne's hungrily consume feed but look thin and scruffy. While diarrhea is common in cows with Johne's, the symptom is seen in goats only at the very end or not at all. Once Johne's is present in a herd, it is

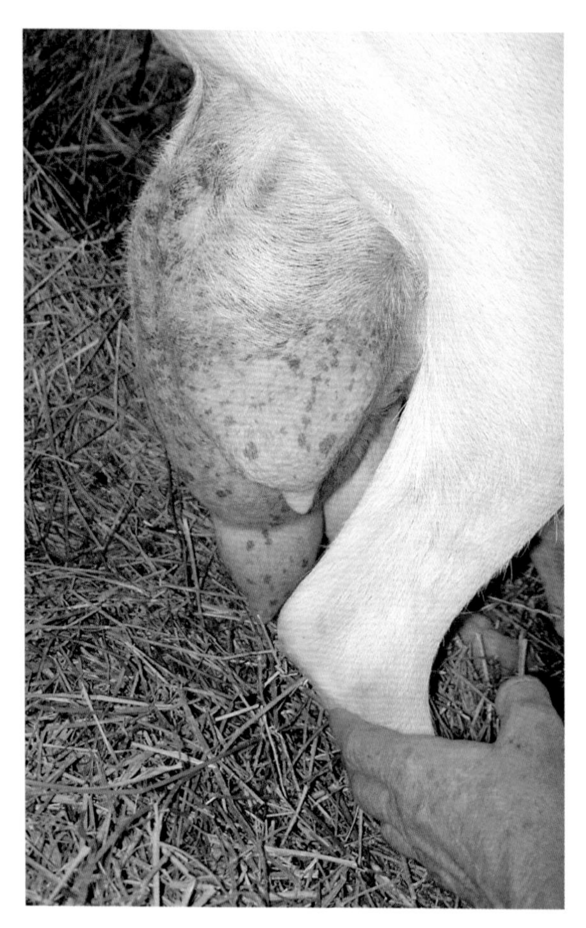

Infections of the udder, called mastitis, can cause deformity and ruin half or all of the milker's productive ability. *Jen Brown*

hard to get rid of. Highly contagious, it is spread by actions as simple as a kid nibbling infected hay. This infection is not treatable and there is no approved vaccine. The best way to prevent Johne's is to use CAE prevention techniques—raising kids separate from the herd.

The Johne's test, along with CAE and CL, is one of the most frequently ordered tests used by breeders claiming disease-free status. Laboratory testing is available for Johne's but is not as reliable as some of the other tests. Negative results can be misleading. Future tests may turn positive without any clear source of infection. Goats who have been vaccinated for CL or have had CL abscesses may give a false positive or borderline Johne's test. It is most reliable as a whole herd screening done over multiple years. To definitively test for Johne's infection, a stool culture or PCA test is recommended.

MASTITIS

Mastitis is inflammation of the udder caused by infection, stress, or physical injury to the udder. It seen most commonly in the dairy barn and in nursing dams. Symptoms include a hot or swollen udder and milk that has lumps, strings, thickening, bad odor or taste, or an off color. If dams are raising kids, the first sign of mastitis may be when the doe refuses to allow the kid to nurse, or the kids appear not to be growing well. Organisms most likely to cause mastitis in goats include *Staphylococcus*, *Streptococcus*, *Pasteurella*, and *E. coli* or other coliforms.

Blue-bag, or gangrene, mastitis is a rapid-onset infection that can be fatal even after immediate treatment. The infected half will be hot and show increasingly blue or black color. This form of mastitis requires veterinary assistance immediately. In some cases, amputation of the affected half is necessary to save the goat. In future freshenings, a goat who has only half a functioning udder can still produce milk from the remaining half.

Prevent mastitis by handling the udder gently and not overmilking. Check milk using a strip cup before each milking as an inexpensive screening. Any lumps, changes in texture, or blood seen in the strip cup should be investigated for possible infection. Wipe dirt or debris from the udder before milking and finish the milking with teat dip or a spray such as Fight Bac. Keep pens and barnyard as dry as possible. Built-up manure or muddy conditions may lead to udder infections.

The California Mastitis Test or CMT screening test for mastitis is used to check suspicious milk. The reagent clumps in the presence of leukocytes (WBCs), forming a visible gel to indicate infection. The test was designed for cows and care should be used when interpreting this test in goats. Goats shed more epithelial cells than cows. Since these cells react similarly to WBCs that occur during infection, a "trace" or "1" result on the cow scale should be considered normal for a goat. A reading of "2" or greater could signal an infection.

A milk culture can be done to determine what organism is causing the mastitis. If treatment will be started before a culture is done, collect a little milk in a sterile container or freezer bag. Save this milk in the freezer or refrigerator to be cultured if the goat gets worse or the original treatment isn't working. Milk cultures collected once treatment has started may not give an accurate organism identification because of antibiotic interference.

Treatment for mastitis is varied. Frequent milking of the infected half relieves discomfort and stress on the udder. Peppermint oil massage alternated with hot packs increases circulation as well as helping to reduce swelling. Teat infusions, either medicated or holistic, are often used to treat mastitis. Some breeders report success with high doses of vitamin C administered orally or infused.

Concentrated injectable antibodies such as Bovi Sera or Goat Serum Concentrate have been used as a natural treatment for mastitis. The recommended two-day protocol is 10cc of the serum injected daily along with 15cc infused into the infected side after each milking.

A dry treatment between lactations will help clear up subclinical mastitis. Suspected animals, or the whole herd as preventative, are infused with a dry-off antibiotic treatment on the final day of milking for the season. Infuse treatment into the udder following the last milking and massage into the udder. Do not milk after that, merely allow the goat to dry off.

NEUROLOGICAL PROBLEMS

Neurological problems are caused by deer worm, listeriosis, polioencephalomalacia (PEM), and any other disease that causes inflammation of the brain. Signs of neurological impairment include staggering, circling, and

Because goats are social animals, infectious diseases like pinkeye or hoof-and-mouth disease can ravage a herd in weeks. Take care to quarantine infected animals apart from healthy ones. *Carol Amundson*

gazing off into the distance or straight up at the sky ("stargazing"). Extreme cases suffer from convulsions or blindness. Treat with massive doses of fortified B vitamins several times a day—and call your vet. Depending on the cause, antibiotics or wormers may be necessary.

PINKEYE OR INFECTIOUS KERATOCONJUNCTIVITIS

Keratoconjunctivitis is inflammation due to dryness of the eye. Some breeders suggest letting the disease run its course without treatment. However, since the problem is easier to treat in the early stages and can cause significant health issues when left untreated, I don't recommend this approach.

In the beginning, keratoconjunctivitis is seen as redness or "pink" eyes. A daily eye check of your goats can catch this problem at an easy to treat point in time. Pinkeye causes red, weeping eyes that can progress to cloudiness and, eventually, blindness. Severe cases develop to a point that ulcers form and can even burst. Before assuming you are dealing with an eye infection, check the eye for dust, hay awns, or other debris that may be causing mechanical injury.

Infectious keratoconjunctivitis is highly contagious. The infective organisms are present in exudate from watery eyes that contaminates the goat's environment. Eye ailments are frequently seen in the summer when flies spread the problem between goats.

Chlamydia infections often present as pinkeye in the initial stages. *Goat Medicine* lists chlamydia and mycoplasma as the most common infective agents causing pinkeye in goats. The pinkeye seen in cattle is not the same as pinkeye in goats, so vaccination with cattle products is not useful.

When an individual goat presents with an eye infection, isolate it from other goats. Using gloves, gently clean the goat's eyes and face with a paper towel dipped in normal saline, a mild Epsom salt solution or alcohol-based mouthwash. Cleaning provides some relief and helps avoid spreading possible infectious agents. In simple cases, optical eye ointments or antibiotic eye sprays applied several times daily will clear up this problem. For an old-time treatment, spray the eye several times daily with port or red wine. Do not use sprays with steroids if the eye has ulcerated, as it can permanently damage the eye.

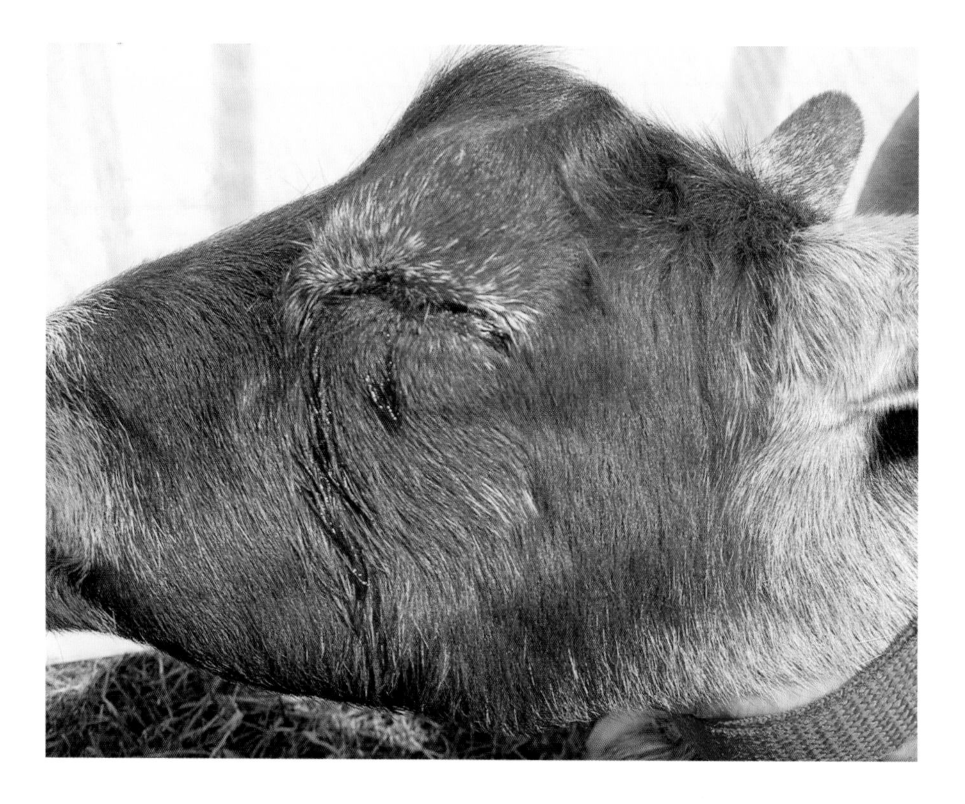

Tearing, cloudy eyes, and blindness are signs of pinkeye. *Jen Brown*

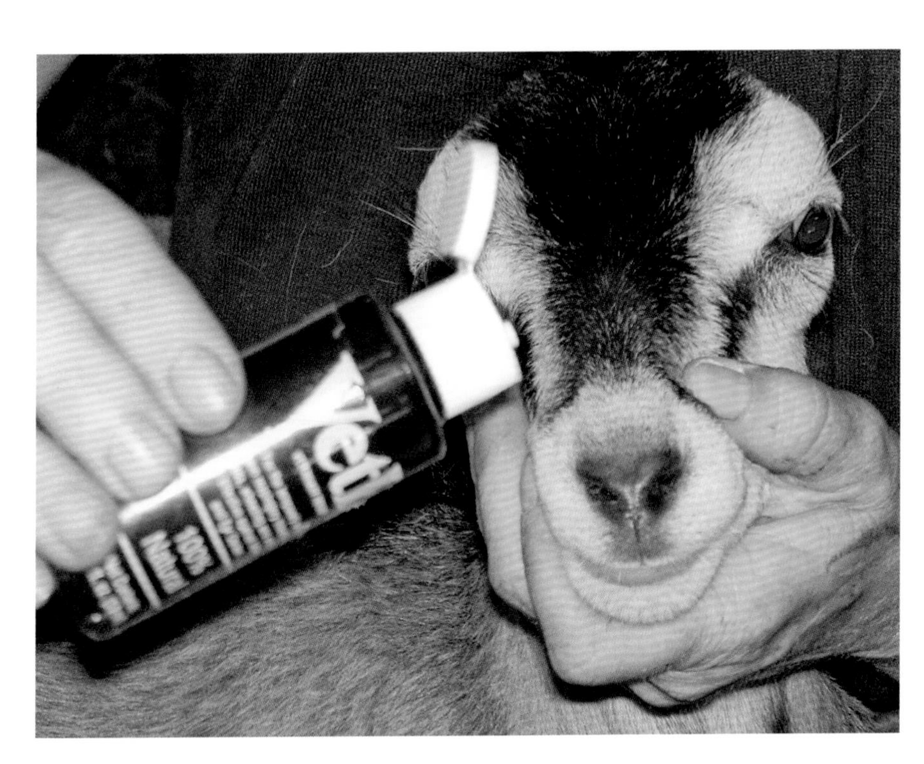

VetRx nose drops are a nonmedicated method of relieving minor sniffles. *David Weber*

When chlamydia is suspected, oxytetracycline such as Biomycin injections and sprays are the treatment of choice. LA200 is used by many veterinarians and owners, but I prefer Biomycin because the carrier is known to be less irritating than the carrier in LA200. Chlamydia is of special concern due to this organism's tendency to cause abortion.

RESPIRATORY ILLNESS AND PNEUMONIA

Respiratory illnesses are caused by both bacteria and viruses that may initially lie dormant in goats and then be triggered by a stressor. Sometimes called "shipping fever," caprine pneumonia is frequently seen in animals moved from one location to another. Windy or wet weather and sudden changes in weather like we often see in Minnesota can also trigger an outbreak. Goats have a harder time regulating internal body temperature than other species, which predisposes them to catching pneumonia.

Interstitial pneumonia is the most common and deadly respiratory disorder in goats. It comes on rapidly and the first sign of a problem is often a dead goat. If your animal isn't dead, the progression of interstitial pneumonia begins with a high temperature followed by a drop in temperature as the lungs fill with fluid. If your goat is standing off to one side, acting depressed or just "off," your best tool is a thermometer. A temperature greater than 104 degrees Fahrenheit without other symptoms can indicate bacterial pneumonia and the goat should be treated immediately. By the time the animal's temperature is 100 degrees or less, survival of the goat is doubtful.

Nasal discharge, coughing, fever, labored breathing or rattling, and depression also signal serious respiratory problems. Consult your veterinarian for treatment. Antibiotics should be used only if a temperature is present. Banamine or baby aspirin will help reduce the fever. Mild cases can be relieved using VetRx, Vicks VapoRub, or other herbal remedies that contain menthol or eucalyptus.

There are pneumonia vaccines for goats, sheep, and cattle. *Mannheimia hemolytica* and *Pasteurella multocida* vaccine by Colorado Serum is labeled for goats. Two other vaccines used by some vets and breeders are Poly Bac B Somnus or Presponse HM. Before using these vaccines, talk to other breeders and your veterinarian to see what is recommended for your area.

RINGWORM

Another disease frequently brought home from goat shows is ringworm, a painful nuisance to the goat. Goats lose hair in scaly, circular patches. Antifungal shampoo or athlete's foot spray are simple, over-the-counter treatments. A goat with a healthy immune system is unlikely to get ringworm after initial exposure. Goats with weak immune systems can have more severe infections or relapses. Take care to use gloves and wash well after contact. Ringworm is zoonotic, so you can catch it from your goat. Trust me—it hurts!

SCRAPIE

Transmissible spongiform encephalopathies (TSEs) are of concern for both the economic loss of sick livestock and the potential for human infection. These conditions are known as mad cow disease in cattle, chronic wasting disease in deer, and scrapie in sheep and goats. Scrapie was recognized in the United States in 1947, in sheep in Michigan. It is far more common in sheep than in goats. Based on slaughterhouse statistics collected through 2016, the incidence in goats is 0.002 percent. Goat owners need to be aware of this disease in spite of its rarity because scrapie regulations for sheep also apply to goats.

Scrapie is a slowly developing neurological disease. Destruction of suspected animals and postmortem examinations are the only way to confirm the disease. If a breeder who keeps both sheep and goats has a diagnosis of scrapie in the sheep herd, all of the sexually intact goats must be destroyed. Breeders with valuable goats might want to consider not keeping sheep or, alternatively, keeping sheep that are genetically resistant to scrapie. Genetic resistance tests are available for both sheep and goats.

The goal in the United States is to gain scrapie-free status, which would increase export markets for both goats and sheep. Only Australia and New Zealand currently have no scrapie in their countries. Scrapie eradication programs require exhibited animals to be from farms belonging to a scrapie ID program.

SORE MOUTH

Sore mouth, also known as orf or contagious ecthyma, is a common but painful condition characterized by blisters, pustules, and open sores on the mouth of a goat. It can also affect the teats, ears, and other places that a kid has tried to nurse. The biggest concerns caused by sore mouth are the failure of kids to eat, secondary infection of the sores, and contamination of the environment. Scabs that fall on the ground can infect the environment for years to come. Sore mouth is transmissible to humans.

Livestock shows prohibit animals with sore mouth. Unfortunately, within two to ten days of returning home from an ostensibly clean show, newly infected animals will start showing signs of disease. As a preventative measure, isolate animals that are new to the farm or returning from shows.

Mostly an annoyance, sore mouth generally goes away on its own within six weeks. Udder balm on infected teats can help reduce pain. Kids that are not eating may need to be tube fed. Antibiotic treatment for secondary infection may be necessary in severe cases.

A vaccine made with live organisms is available. Use it only if your herd has had the disease; otherwise you're likely to bring it onto the premises by the very means intended to prevent it.

TETANUS

Tetanus is an often-fatal disease caused by a bacterial toxin. Animals and people contract tetanus through a deep wound or cut that doesn't bleed properly. Castration and disbudding may cause infection from tetanus. Also called lockjaw, its symptoms include spasms, loss of coordination, and stiffness, especially in the jaw.

Prevention is always better than treatment, and this is especially true for tetanus. It is strongly recommended that you vaccinate your goats with CD/T, which gives immunity to enterotoxemia C and D and tetanus. In young kids from unvaccinated mothers, give the tetanus antitoxin for short-term protection.

The prognosis is not good for a goat that has contracted tetanus and started to stiffen. Treat with large doses of penicillin and tetanus antitoxin. Holistic practitioners believe that vitamin C detoxifies tetanus victims.

TUBERCULOSIS

Tuberculosis has not been found in goats in the United States for many years, but it continues to be monitored due to its prevalence in cattle. A simple skin test can be done by a vet to see if a goat has had contact with the disease. This skin test is often required for interstate shipping or export.

URINARY CALCULI

Urinary calculi are tiny stones or crystals that form in a goat's urinary tract, causing pain and blockage. Bucks and wethers are most susceptible to this condition, which is also known as "water belly." Symptoms include straining to urinate, dribbling, blood in the urine, or pain exhibited by grinding teeth or lying down and kicking at the abdomen. Many cat owners are familiar with this problem. A blocked urethra is life-threatening. Call a vet if blockage is suspected. Sometimes, a simple removal of the tip of the urethra will relieve the blockage.

Prevention is the best medicine. Feed a ration with a calcium-phosphorus ratio of at least 2:1, never lower than 1:1. Don't add phosphorus or magnesium to feed or allow the goats to gorge on concentrated grain rations. Overconditioned animals, especially wethers, are more susceptible. Prevent calculi by providing plenty of fresh drinking water and supplementing the diet with ammonium chloride. Many commercial goat feeds already contain ammonium chloride.

Goats with sore mouth may go off their feed, leading to secondary problems. It is highly contagious. *Jen Brown*

5 LIFE WITH GOATS

IF THE OWNER OF THE GOAT IS NOT AFRAID TO TRAVEL BY NIGHT, THE OWNER OF A HYENA CERTAINLY WILL NOT BE.

—NIGERIAN PROVERB

Most goats are excellent travelers. Because of their small size, goats can be transported in many different vehicles. The key factors are the same as in housing: avoid wind, extreme temperatures, and wet conditions. In trailers or carriers, the sides should be smooth with no protrusions or sharp edges. No carrier should be totally enclosed, because poor ventilation can cause stress and possibly suffocation, especially in hot weather. Exhaust fumes from other vehicles seeping into poorly ventilated spaces can cause carbon monoxide poisoning.

Use common sense when moving your animals. Those that are sick, lame, or heavily pregnant need special care. These goats should be transported only under conditions of need, such as veterinary care. Tying a goat in the back of an open vehicle could lead to injury or strangling. Tying a ruminant so that it's lying down on its side can cause bloat.

Standard transport for goats is either the bed of a covered pickup or a trailer. Use bedding, such as straw, unless you are going a short distance. Bedding provides a comfortable

Above: Goats often love to go traveling once familiar with the vehicle. Our goats gather around the trailer because they enjoy traveling to shows. *Jen Brown*

surface and soaks up animal waste. Slippery floors can cause goats to fall.

When loading a truck or trailer, make sure there is enough room for each animal to move comfortably. The FAO recommends 4.5 square feet per goat. In an overloaded trailer, a fallen goat can be trampled. When driving a truck or trailer with animals in the back, drive smoothly, avoiding jerks or sudden stops. Make turns gently.

Feed and water your goats before a long trip to keep them more comfortable. Goats need feed and water every twenty-four hours at minimum on long journeys, and it is best if they can be offloaded to stretch their legs. This rule is applied to meat animals being shipped for slaughter but also pertains to goats moving cross-country for shows or to a new home.

Mixing horned and hornless animals is risky. The same is true for mixing kids and adults or bucks and does. It helps if the animals are familiar with one another so that they won't have to settle questions of rank in the confines of a truck. Naturally, a randy buck or several bucks with does is a recipe for disaster. I mix adults and kids only when the truck is partially full and the kids are traveling with their dam.

Small-breed goats or kids fit nicely in travel kennels used for dogs. The larger goat breeds can travel in similar boxes that have been specially made to fit in a truck or car. There are even cages designed to go in the bed of a pickup for short rides in nice weather.

Just like dogs, goats will ride comfortably on the back seat. It is wise to have a mat or blanket and a well-mannered, tame animal when trying this method of transport.

INTERSTATE SHIPPING

Myriad rules cover the transport of animals across state lines. Usually, a minimum requirement is an interstate Certificate of Veterinary Inspection. Show rules usually spell out the conditions for out-of-state exhibitors. Find out what requirements are needed by your destination state and any states you are traveling across. A good resource is the USDA Animal and Plant Health Inspection Service. The USDA website has links for individual state and international regulations.

A homemade wooden box for the bed of a pickup can be all you need for transport. *Jen Brown*

FLYING A GOAT

Most goats that travel by plane are kids purchased for replacement or breeding stock. A small goat breed or a kid can often be shipped as a pet. Different airlines have different rules, but all are governed by Federal Aviation Administration and USDA standards. Some basic guidelines include the following: Provide a standard dog crate large enough for the goat to stand up and turn around. Attach food and water to the crate. Provide a collar and leash for ease of inspection. When shipping goats in cold weather, provide a veterinary acclimation statement saying that the animal can withstand the temperature extreme. Do not ship goats in hot weather. Be prepared for the inevitable airline searches, delays, and mix-ups.

INTERNATIONAL EXPORT

Export rules have gotten particularly strict in light of biosecurity concerns. A number of borders have been closed to goats from the United States, including all of Europe. Other

With a nice blanket, the back seat is comfortable! *Jen Brown*

Getting into a pickup can be a tall jump for the goat. Some goats need a little help getting into the truck. We have others that jump right in. *Jen Brown*

countries have explicit testing or quarantine regulations. Most owners who sell goats into another country work through a buyer who knows the needs of the customer. In some cases, the intermediary purchases the goat and performs the testing and quarantining.

In July of 2018, the USDA reported that plans are being finalized for sheep and goat imports to resume in Japan. The shipping of goats into Japan has not been allowed for fourteen years. This is very good news. In the 1990s, I exported goats to Japan, Korea, and elsewhere. As a small breeder, it was a welcome source of income.

RECORD-KEEPING

Various regulations, as well as the proposed National Animal Identification System, require breeders to track the movement of their livestock. Keep records of the tag or ID numbers of any animals permanently leaving your property, the dates they left, and their destination. Similar records should be maintained for purchased goats.

GOAT HANDLING

Livestock behavioral science is a field of study that helps owners manage their animals in ways that are safe for both handler and livestock. Many goat owners spend quite a bit of time with their charges. In contrast, goats on range may only be handled seasonally. An understanding of caprine behavior along with some advance planning will make chores more enjoyable for both humans and goats.

General livestock handling has a series of useful principles for understanding how to manage your goats. These are the four principles:

- Position—handlers should work from the side, not the front, and avoid the blind spot at the rear
- Pressure—move closer and the goat moves away, back up and the goat stops or moves toward the handler
- Movement—increase or decrease handler movements to get animals in place, which often involves broad arm motions and can involve tools like sticks (which should not be used to hit animals, only to signal motion)
- Communication—the handler clearly lets goats and other handlers know what they need

Goats on range that are not used to frequent contact with people will not allow handlers to catch or touch them easily. These animals need more specialized ramps, pens, and handling chutes (also called races) than tame goats do.

Some of this behavior can be described as the "flight zone," or the goat's personal space. Goats that are around people all the time, such as dairy goats, may have little or no flight zone or their flight zone may be different in the parlor than in pasture. When strange people or animals are brought into a situation, the flight zone will change based on the goat's experience. Approaching a goat from the front increases the flight zone; the zone is smaller when the animal is approached from the side.

Goats are more difficult to handle than cattle or sheep. They stress more easily and can become aggressive toward each other when crowded. Frightened goats will pack into corners or even lie down and sulk. When together in a familiar herd, family groups, usually led by a matriarch, cluster and move together. Goats observe a herd pecking order. While milking at Poplar Hill, even when we had a herd of five hundred milkers, the animals usually came into the parlor in roughly the same order every milking based, on a system that was clear only to the goats. It is always best not to try and force the order.

Prey animals like goats are most comfortable with their herd around them. This same behavioral trait will keep a goat following its owner or herd if you want to move it from one place to another. Goats do not like to be alone and will run back to the herd when given an opportunity.

Goats can see 320 to 340 degrees, a panoramic view of their world. There is a 40- to 60-degree blind spot at the rear of the goat behind its shoulder blades. If a handler stands in the blind spot, the goat will turn to face the person in the blind spot and stop a line of goats from moving. When handling goats in chutes, the wide vision may act to distract or panic the goat if strange people or movements can be seen through the sides. Solid sides on chutes can reduce escape attempts or goat panic attacks.

When possible, holding areas or milking parlors should have an entrance and an exit to keep an easy flow and prevent jam-ups of animals trying to move both directions at once. A chute or race should ideally be only the width of a single animal to prevent animals from turning around.

Unless it's an emergency, goats should not be lifted off the ground or dragged by a single leg or the head, ears, horns, neck, tail, or fleece. Grabbing a goat horn above the base is dangerous. The top of the horn can break, causing a bloody injury. Grabbing a goat by a limb can cause disarticulation of the joint.

HERD DOGS

If you are planning to use herd dogs to work with the goats, be aware that goats are more difficult to herd than sheep or cattle. It is best to start with a more experienced dog instead of a pup.

My friend Marie had a dog called Deacon, a very well-trained Sheltie with advanced placements in obedience trials. Deacon "retired" to my farm. I was looking forward to using him to sit in the goat aisle to block the milkers from going the wrong direction after milking. The first evening I brought the poor dog out for chores, the goats ran him right over. He turned tail in the aisle and raced from the barn. I made a novice stock-handling mistake by not asking about Deacon's experience. Beaten up by Marie's goats when he was a pup, Deacon was terrified of the species. He'd herd sheep just fine but wouldn't ever work the goats. And he refused to visit the barn all his remaining life!

GOAT CHARACTERISTICS AND BEHAVIOR

- Have excellent peripheral vision, with blind spots just behind shoulders
- Have great distance vision, with some problems judging distance accurately
- Can distinguish colors—sensitive to yellow-green and blue light
- Avoid the dark (shadows will cause goats to balk)
- Tend to move toward light, unless it is shining directly in the eyes
- Prefer to follow a leader
- Can become stressed when separated from the herd
- Will move in a circle around the handler or pen
- Have a keen sense of hearing
- Are easily distracted by noises or sudden movements
- Prefer routine—goats are creatures of habit
- Have good memories and will remember bad experiences for a year or more
- Don't like to walk in water or deep mud
- May jump high or attempt to jam through barriers to escape

Though goats are herd animals, they will split up more often than sheep. A herd dog expects that the stock will move in a group when the dog makes eye contact, but stubborn goats will often stand their ground and stare back. My LaManchas will turn to face a dog, stare into its eyes, and even lower their heads to butt. Bucks, herd leaders, or more assertive goats challenge dogs most frequently. At the same time, the shyer goats, usually my Nubians, will "popcorn" off in all directions to get away from the canine intruder.

When they run, goats don't just run on flat ground. Goats can also jump up on things like the picnic table, a car, round bales, and even a hut or other building. Dogs unused to goats may be confused by these tactics. Take care that frustrated or ill-trained dogs aren't harrying the animals or biting.

KEEPING MILK GOATS

The dairy products of goats—milk, cheese, yogurt, sour cream—are a healthy, tasty alternative to cow products. People accustomed to commercial cow milk will find goat milk richer, slightly sweeter, and varying in flavor with the season. Commercial cow herds contain animals in various stages of lactation. By the time their milk is shipped, processed, and packaged, its butterfat percentage has been adjusted and the flavors standardized. Fresh goat milk is as comparable to this product as fresh vine-ripened garden tomatoes are to the tasteless ones from the supermarket.

I love giving two glasses of milk to someone unfamiliar with this tasty product. One glass has whole cow milk, the other goat milk. When asked, "Which is the cow milk?" the person holds up the goat-milk glass. Why? "Because it tastes better."

As a health food, goat milk has a lot going for it. Although it is higher in fat than cow milk, the fats contain more omega-3 fatty acids, the same ones in fish oils touted for increasing "good" cholesterol. The fat molecules are also

Artisan chèvre cheese is made from fresh goat's milk. *Shutterstock*

smaller, so they stay suspended longer than fat molecules from cow milk. This means that goat milk is naturally homogenized and only separates very slowly. You won't see a lot of homemakers using goat milk butter or whipping cream for this reason.

Many people become interested in raising goats because they have allergies or other health issues that prevent them from drinking commercial cow milk. They search out a local farm to provide goat milk on a regular basis, and soon they want the milk source at hand. That's when they buy a couple of dairy goats—and the fun starts.

HOME VERSUS COMMERCIAL DAIRYING

Home dairies operate at the whim of the owner. As long as the milking parlor is clean and the milk handling correct, no other rules apply. The goat owner who chooses to go commercial has an entirely different operation. Any commercial dairy must follow the regulations of its home state. Each dairy is licensed by its state and pays an inspection fee at least once each year. Interstate shipping regulations require additional inspections, licensing, and fees. Inspections cover everything from the milk holding temperature to feed storage to veterinary treatment.

I FIND AMONG WRITERS THAT THE MILK OF THE GOAT IS NEXT IN ESTIMATION TO THAT OF WOMEN, FOR THAT IT HELPETH THE STOMACH.

—WILLIAM HARRISON, ELIZABETHAN ENGLAND (1577)

MILK HANDLING

Milk is a wonderful food for goats and humans—and for microorganisms. It is impossible to produce completely organism-free milk, but microorganisms must be controlled. Good microorganisms help make yogurt and cheese and also make milk more digestible. Opportunistic microorganisms, though, can spoil the milk at the very least and cause illness in a worst-case scenario, such as infection with *E. coli* or *staphylococcus*.

To keep milk from spoiling, everything that touches the milk must be sterile. In the simplest terms, clean milk directly from the teat is filtered into a sterilized container. The first squirts of milk should be collected in a strip cup. Throw away the milk in the cup because it could contain dirt and contaminants from the end of the teats. Sterile procedures must be followed whenever milk is transferred to other holding containers.

Ideally, a milk pail should be stainless steel with a lid. For commercial dairies, stainless steel, food-grade plastic, or clear glass are the only nonfilter materials allowed to contact the milk. For private use, any container that can be bleached is fine.

In a blind taste test, people unfamiliar with goat milk often find it tastier than cow milk. *Shutterstock*

Bleach is the best sanitizer for dairy equipment. Once the container or other implement has been washed, use a hot-water rinse of 2 percent chlorine sanitizer or bleach. Submerse the item in the sanitizer bath, then hang it or set it on a clean rack to drip-dry.

Around 1862, the French chemist Louis Pasteur discovered that heating liquid killed germs. This discovery revolutionized food handling and safety. The spread of tuberculosis and brucellosis through contaminated milk has since been vastly reduced. For proper pasteurization, milk should be heated to 161 degrees Fahrenheit for fifteen seconds. Heated milk should be cooled rapidly to 36 degrees in an ice or cold-water bath and placed in refrigerated storage.

Raw milk is controversial. While I enjoy drinking our milk raw, I know the health status of my does and the handling of my milk. The American Veterinary Medical Association stands firmly on the side of pasteurization. In many states, it is illegal to sell raw milk.

Consumption of raw milk should be an informed decision. Do not use raw milk for infants or people who have compromised immune systems. I pasteurized milk for our daughter until she was two years old as a precaution against illness.

Off-Flavored Milk

Normally, fresh goat milk is sweet with no strong or "goaty" flavor. If your goat produces unpleasantly strong-flavored milk, check your handling process. The longer milk is held in storage, the more the fatty acids break down, producing a stronger flavor. The higher the butterfat, the slower this breakdown seems to take. For this reason, the higher butterfat Nubian or Nigerian Dwarf goat milk is often reported to be sweeter and keep for a longer time than milk from other breeds.

The food your goat eats also affects her milk. Plants like ragweed or goldenrod may flavor the milk. Strong kitchen scraps can have an effect, particularly onions, garlic, cabbage, broccoli, grape leaves, or blackberries. The same principle applies if she is deficient in cobalt, vitamin B12, vitamin E, or protein.

The milking parlor is where the actual milking of the goats occurs (right). The processing of the milk occurs in a separate room called the milk house (right, below). *Terrapin Acres*

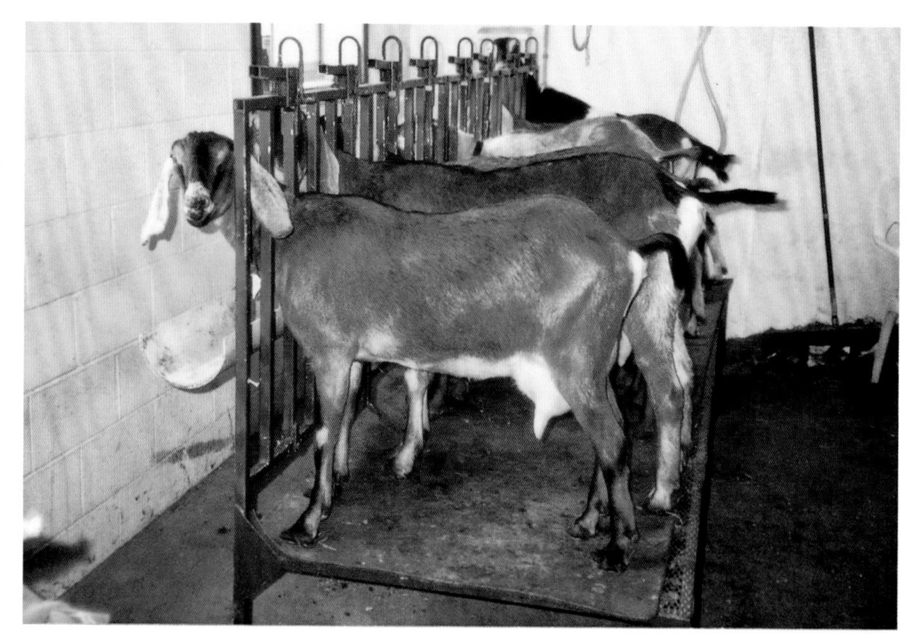

Some people claim that keeping bucks with your does can cause off-flavored milk. That has not been my experience. However, some does may have stronger tasting milk when their hormones surge during heat. Breed or lineage is another factor sometimes listed for strong-tasting milk. Most often mentioned in this regard are Toggenburgs or Oberhaslis. When buying an adult milker, it can help to taste fresh milk from the doe before taking her home.

Mastitis

In some cases, off-flavor milk is due to mastitis. The milk may taste a little salty, or it may develop an off-color, stringy texture and bad smell. Lumps or strings can be detected by squirting the first streams of milk from each teat into a screened strip cup. During late lactation, mastitis is commonly subclinical, meaning that it's not noticeable except through testing. Commercial dairies regularly test milk for the organisms that cause mastitis. Blood in the milk may indicate infection. However, in early lactation, pink color is often due to the breaking of tiny blood vessels in the udder as it grows to accommodate the milk. Investigate any bluish color to the udder, excessive hot or cold feel to the skin, or sudden hardness or unevenness. Do not drink milk from infected goats. For more on mastitis, see page 156.

THE DAIRY BARN

The dairy barn should be set up for convenience. Have water-access, milk-handling, and clean-up implements and supplies nearby for the home dairy. Commercial regulations specify everything from the way water is delivered into the area to what materials may be used in the walls. In a formal milking parlor, the walls and floor are made of a material that can be washed. Flies must be controlled.

Milk stands are readily available in many styles. My first milk stand was homemade. Later, when I had my commercial dairy, I had double metal stands to hold two goats. These stands can be placed side by side, creating a row of milkers, yet they are easily moved for cleaning. A portable milk stand can be taken to shows. Concrete stands work well in a formal parlor with metal head gates built in. The concrete is easily swept or washed.

Commonly, each milk stand has an attached feeding trough or bin. This setup ensures that the goat is getting her feed and keeps her busy while you work.

Plan how the goats will come into the parlor. If you have only a couple of goats, they can just jump up on a stand. As the number of goats increases, however, you'll need an orderly way for the goats to enter and exit. Some dairies accomplish this with a pen system that lets goats enter by one door and exit by another.

Additional equipment setup will depend on whether you milk by hand or by machine. Each procedure has its advantages and disadvantages. Hand milking requires low-cost equipment and is a peaceful way to spend time twice a day. Using your hands can also help you monitor udder health. Many goat owners refuse to milk by machine simply because of the mastitis risks involved with machine milking.

Machine milking has a high initial setup cost and requires technical knowledge. If the equipment is set up and used properly, however, machine milking allows for less experienced farmhands to help with chores. While some machines can be loud, locating the compressor outside the parlor eliminates much of the noise. Once goats become used to the machine, most will allow even strangers to milk them without complaint.

Milk handling and processing are usually completed in a room separate from the milking parlor. The milk house usually has water hook-up and cold storage. For the small home dairy, these steps are readily handled in the kitchen. Having a separate room close to the milking parlor for processing milk, especially for feeding milk back to kids, keeps the clutter of utensils and the mess of spilled milk away from the house.

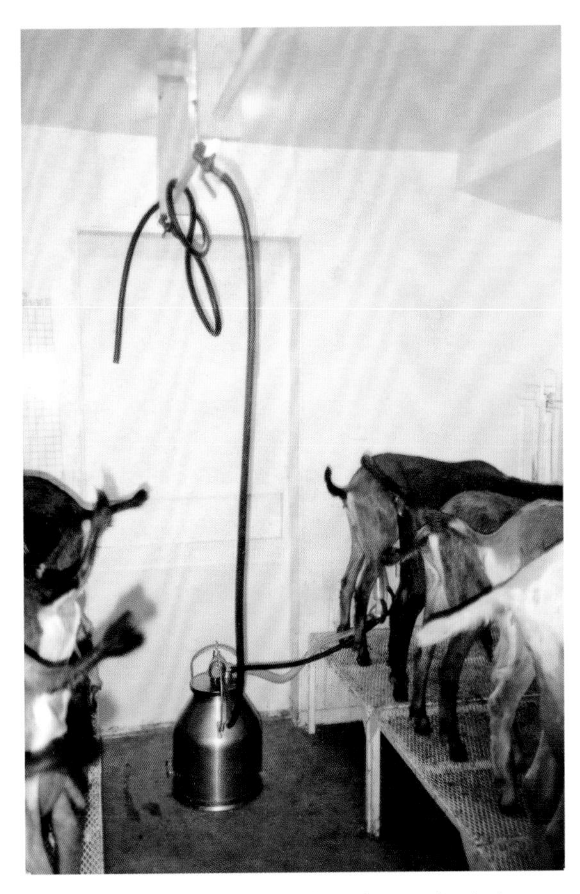

A formal milking parlor has metal head gates, elevated stands, and washable floors and walls. *Terrapin Acres*

HAND MILKING

By far the most common way to get milk from a goat is by hand milking. After a little practice, milking can become a great way to spend time with your goats. An adept milker can milk a goat out in three to five minutes. Do not be discouraged if it takes you longer at first. My first experience took twenty minutes, and I got more milk on me than I did in the pail—and the milk was brown because the goat's manure-covered feet got into the pail as well!

Goats can learn to be milked from either side or from the rear. Most often, milkers work from the side. I milk from behind because it is easier on my back, and I like being able to see the whole udder to check its condition. Unlike cows, goats usually don't kick backward and rarely urinate or defecate while being milked.

Pigs love goat milk. Raising other livestock on goat milk can be fun and makes use of excess milk on noncommercial dairy farms. *Terrapin Acres*

HAND-MILKING EQUIPMENT

- Acid detergent
- Bleach or sanitizer
- Clean-up brushes
- CMT kit or mastitis indicators
- Dish soap or dairy soap
- Milking stand
- Paper towels or dairy towels
- Stainless-steel milking pail
- Stainless-steel strainer and milk filters
- Strip cup
- Teat dip, such as Fight Bac

When milking from behind, the handler can always sit without twisting—and see the entire udder. *Carol Amundson*

Milking from either side is more common. To prevent startling the goat, it helps to talk to her and let her know you are going to touch her. *Carol Amundson*

The handler should grasp the udder gently but firmly. Allow the teat to fill with milk, then close off the top of the teat with your thumb and forefinger. *Carol Amundson*

Squeeze the rest of the teat with the other fingers until the teat has emptied. Release the top of the teat and repeat the process until the udder is empty. *Carol Amundson*

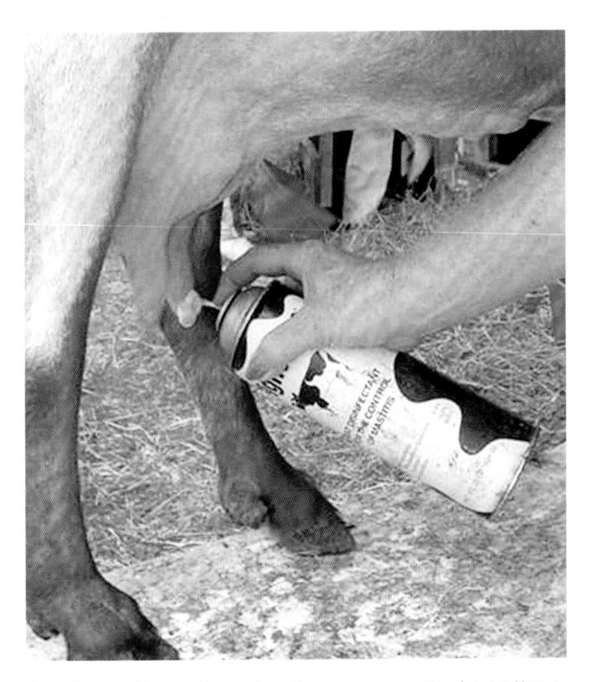

Spraying or dipping the teat with an udder wash after milking will help protect the doe from mastitis until the teat end closes off. *Carol Amundson*

Cabrito is the tender meat of a milk-fed kid slaughtered at 20 to 35 pounds. It is popular with the Hispanic market for Cinco de Mayo and other celebrations. *Shutterstock*

KEEPING MEAT GOATS

Cabrito, capra, capretto, chevon, chivoor, just plain goat meat—whatever the name, meat from goats is a staple in many cultures. It is popular in Mexican, Greek, African, and Arabic cuisines. Goat meat is favored because there are few religious restrictions toward eating goat. Several traditional celebrations feature goat. In the United States, most large-scale meat-goat operations can be found in the South, where ethnic markets create a consistent demand. That said, most of the goat meat consumed in the United States is imported due to the lack of processing centers to meet the US demand. Those imports, largely from Australia, continue to grow. Australia saw its export of goat meat jump 38 percent in 2017.

I previously quoted a falsehood that goat meat is the most consumed red meat across the world. Sheep and goat meat are counted all in one category, falling behind pork, poultry, and beef. Across the world, however, goat meat is consumed by 63 percent of the population. Prior to 1999, the United States was a net exporter of goat meat. Within the past decade, goat meat consumption in the United States changed dramatically and the United States became a net importer of goat meat.

Though goat is building in popularity as a specialty meat product, the average American has still not eaten it. Once they try it, many people enjoy goat meat, finding it indistinguishable from other red meats. Every ten years for the past three decades, the number of goats slaughtered in the United States has doubled. Numbers continue to change as most of the slaughter counts come from inspected plants rather than personal or unregulated slaughter. In 2008, the US Goat Survey started to count non federally inspected slaughter. For the years 2006 and onward, the number of slaughter goats increased by more than two hundred thousand head each year by adding these animals to the census.

The increase in numbers is attributed to many factors, including the growth of ethnic cooking, interest in nutritional content of meat, the sustainability of goat production compared to other livestock, and food trends often led by the popularity of televised cooking shows and epicurean blogs. All in all, the expectation is to see goat meat rise in popularity in the same way that goat cheese became accepted twenty years ago.

Food choices are changing as people become conscious of the environmental and health impacts of what they eat. Consumers are looking for more grass-fed meat raised under humane conditions. Some are seeking local products direct from the farm. Local providers suit the goat market very well. Some farms lease goats for brush and fire control, an environmentally conscious use of meat-producing animals.

Goat meat also offers health benefits. Roasted chevon has the same number of calories as chicken but more minerals and less fat. It also has the same protein content as beef but 10 percent more iron. Goat meat from grass-fed animals is higher in omega-3 fatty acids (good fats).

These Boer goats are being fattened for market. *Shutterstock*

Above and right: Don't be tempted to fatten your meat goats with a grain-only diet. Even as you are "finishing" your goats for market, they need roughage in their diet. *Carol Amundson (top), Jen Brown (above)*

CARCASS CHARACTERISTICS

A goat generally dresses out at 50 percent or less, meaning that half of the live weight of the animal is removed from the edible carcass. Some of the removed parts are useful in other ways, such as the hide, which can be used for goatskin leather. The older the animal, the higher the percentage of fat in the carcass and the lower the percentage of bone. The largest muscle (and therefore meat) mass is found in the leg and the shoulder. Some people are very disappointed in the amount of meat they pick up from the butcher following processing. The more boneless cuts in your slaughter order, the less product is obtained, as bone is discarded rather than kept.

FEEDING

Market goats need the same minerals and vitamins as other goats but differ in protein and energy requirements. Feeding the market animal during the last few months of growth is known as "finishing." Finishing the goats solely on pasture causes slower growth compared to a feedlot but may be more economical. The meat-goat wether is usually fed a combination of forage, grain and/or protein supplements, and minerals.

Attempting to feed only grain to add weight quickly is a bad idea. Goats can become bloated or develop enterotoxemia from an imbalance of rumen flora without the proper amount of roughage in their diet. One way to avoid overfeeding grain is to give the goat a specific amount of feed and allow fifteen minutes for eating. Remove and measure the amount remaining. Adjust your ration to that level for the next feeding.

RATE OF GAIN

It is important to understand how your goat is growing. At the very least, you should weigh your goat at birth, weaning, and sixteen to eighteen weeks. You can then calculate the rate of gain:

$$\frac{\text{(current weight - start weight)}}{\text{number of days}} = \text{daily weight gain}$$

Example: For a goat that weighed 43 pounds at weaning and 75 pounds after 90 days of finishing:

$$\frac{\text{(75 lbs. - 43 lbs.)}}{\text{90 days}} = \frac{\text{0.36 lbs.}}{\text{day}}$$

Goats being finished should gain about ¼ pound to a ½ pound each day. Monitoring the rate of gain and adjusting feed accordingly will help your goats reach a particular market weight in a specified amount of time.

Some breeders give goats other feed additives in a ration to maintain health and promote growth. Common feed additives include the following:

- Ammonium chloride (helps prevent urinary calculi)
- Baking soda (buffers the rumen)
- Coccidiostat, such as decoquinate (Decco), Rumensin, or Bovatec (prevents and controls coccidiosis)
- Dry mineral blocks (provide specific minerals and vitamins)
- Probiotics (help rumen function and improve feed utilization)
- Wet molasses blocks (provide extra vitamins, minerals, and energy)

MARKETS

Sending your meat goats to a stockyard or an auction is only one method of delivering your product to the consumer. Live animal markets allow buyers to view and purchase goats, which are then slaughtered at an onsite facility. Direct marketing from your farm gives you a chance to meet your customer and develop an ongoing relationship.

The legality of having the buyer visit your farm to select a live animal for slaughtering on the premises depends on your local ordinances. My preferred method is to deliver goats to a local butcher. Also known as "freezer trade," this method saves time and trouble. The consumer picks up the goat already processed into retail cuts.

In our area, as well as other parts of the country, the number of inspected local butchering operations has been steadily decreasing. Loss of slaughter operations limits the ability of producers in affected parts of the United States to offer their product at a competitive price. Transporting animals for several hours to a slaughter facility further costs producers hours of time, wear and tear on equipment, fuel, and carcass losses from stress on the animals.

Ethnic Markets

Slaughter methods are not necessarily considered by the consumer who buys neat, cellophane-wrapped packages from the grocery store. However, slaughter methods do matter to many of the ethnic and religious groups that eat goat meat. Whether the slaughter is handled privately or at a federally inspected plant, knowing the requirements of various groups can make or break your sale. For example, while federal inspections require stunning animals before slaughter, there is a religious exemption in the case of the ritual slaughter of goats for some ethnic groups.

If you sell goat meat to traditional Muslims, you will need to follow strict slaughtering procedures, called halal (Arabic for "lawful"). This includes a blessing followed by zabiha, a humane killing by means of a quick cut across the throat so that the blood may drain. You must avoid contamination with pork products, which are considered unclean.

Shechita is the ritual slaughter of meat animals following Orthodox Jewish tradition. The shochet is the person who performs the ritual, which requires that a knife be drawn across the goat's throat. For kosher consumption, the blood, various veins, fats, and the sciatic nerve must be removed from the meat. The sciatic nerve is difficult to remove from the hindquarter. For this reason, often just the forequarter is sold as kosher.

Timing is also a consideration for ethnic sales. Easter, Christmas, New Year, and other holidays increase the demand for goat meat. For certain holidays or cultures, the weight, gender, and age of the animal are quite specific. Roman Easter celebrants prefer a milk-fed kid weighing around 30 pounds, while Eastern or Orthodox Easter believers seek a slightly larger, milk-fed kid of 35 to 40 pounds.

A lunar calendar is useful for noting Muslim holidays. Ramadan occurs in the ninth month of the Islamic calendar and includes family feasting at the start and end of the holiday. No food is eaten between sunrise and sunset during this month-long event, so the evening meal is frequently something special. Buyers prefer weaned kids weighing around 60 pounds. Eid al-Adha, also known as the Festival of Sacrifice, occurs about seventy days after Ramadan and is a good market for yearlings or large kids weighing 60 to 100 pounds. This animal needs to be good size when possible, since traditionally, one third is given to the poor and needy, while another third is given to relatives. Some buyers want an uncastrated buck.

Goats of 60 to 80 pounds are sought for Caribbean holidays. Older bucks are preferred. The meat is sometimes prepared with the skin still attached and the hair singed away. The goat meat is often cooked for a long time in spicy, stewlike dishes. Chinese buyers like 60- to 80-pound goats. The demand from Chinese buyers is most common in the colder months of the year.

The Hispanic market covets tender cabrito, the roasted flesh of milk-fed kids weighing 20 to 35 pounds. Larger goats are popular for dishes such as Seco de Chivo (goat stew) and whole, pit-barbecued goat for celebrations. Cinco de Mayo (May 5) is a popular day for goat dishes.

There is a meat market in the Hindu community during the fall holiday Navratri, also called Dasara. The main dish served at this holiday is curried goat. The event honors the goddess Durga, and the goat served must be male. Weaned wethers are preferred.

Tennessee Fainting goats are bred for rapid growth and double muscling, making them a good meat breed. *Jen Brown*

KEEPING FIBER GOATS

Goats raised primarily for their hair coat are known as fiber goats. They produce cashmere or mohair fiber for handcrafts and spinning. The highest-value commercial fibers are normally white, because the uniformity of color is useful in mass manufacturing. Colored fleece is valued by hand-spinners and artisans who appreciate the natural variation in tone and hue.

Angora goats are referenced on Sumerian cuneiform tablets and in the Bible and probably evolved between the twelfth and fifteenth century BC. The name Angora comes from the capital city of Ankara, Turkey, where these goats were established in herds. Mohair became a major trade good, with Turkish exports of mohair fabric to Europe in the fifteenth and sixteenth centuries exceeding demand. The Arabic word *mukhayya*, from which we get the name *mohair*, means "selected or choice." This preferred fiber and the animals themselves were embargoed by the Sultan of Turkey. Interestingly, this embargo may not have been absolutely necessary, since the climate is very special in the lands that developed the Angora goat and is believed to impact the softness and quality of the fiber in goats and other animal species.

Sometimes called "living stuffed animals," the Angora goat can give 25 percent of its body weight annually in fiber. Many Angora goats are used by hobby farmers and fiber enthusiasts. Angora goats also fill a commercial niche in the United States. Large commercial herds—predominantly in the Southwest—produced more than 1 million pounds of mohair in 2010. Census figures show that 172,000 animals were clipped, yielding an average of 6 pounds of fiber per goat. The mohair sold for an average of $3.49 per pound.

Culled dairy or fiber goats of any breed can become meat goats for extra profit. *Jen Brown*

RAISING MARKET WETHERS FOR SHOW

Raising a market wether is a popular youth project. The kid is purchased, weaned, and fed to market size for a period of months before being shown and ultimately sold. The time involved is often shorter—and more manageable—than a dairy project.

Check your show requirements before purchasing your goat. Target dates for age, called the "tooth rule," state how many mature teeth the goat may have at the time of show. There are often rules about weight limits, clipping, horns, and veterinary inspection. Usually, there is a validation date for entering your animal, so be prepared to get all paperwork in on time.

An animal project is a commitment. The best way to raise a good market goat is to commit to care every day, twice a day. Often, these animals are pampered almost to death in an effort to produce a rapid rate of weight gain. That isn't really necessary. The goal is a well-muscled and prime goat. Excess fat isn't much better than an animal that is too thin. Feeding standard goat rations, good hay, and minerals in proper amounts, along with implementing a good parasite prevention program, will go the furthest in producing a good-looking, healthy animal.

Cashmere goats, on the other hand, are not counted in census surveys. A cashmere-type fiber can be found on almost any breed, except Angora. In the early 1970s, researchers began investigating the undercoat produced by American goats. They found that many goats did produce a cashmere-type undercoat but not enough to make it economically viable. Through selective breeding, we now have cashmere-producing goats in the United States.

Cashmere is warm, soft, and long-lasting. Because only 4 to 6 ounces of this fine undercoat are produced each year per goat, products made from cashmere are highly valuable. The price of cashmere products varies based on the quality of the fiber and the workmanship of the final product.

The Cashgora goat is a cross between Angora and Cashmere breeds. Angora crosses with miniature breeds are more likely to be found in hobby and pet homes rather than commercial herds. An Angora–Nigerian cross is called a Nigora, and an Angora–Pygmy cross is called a Pygora.

Angora goats produce mohair fiber. Pure white hair often commands the highest price. *Shutterstock*

Created breeds such as this Pygora goat are popular with artisans who like unique fibers. *Fran Bishop, Rainbow Spring Acres, Pygora Breeders Association*

Cashmere goats produce fine cashmere fiber for clothing and other products. *Shutterstock*

BUYING FIBER GOATS

As noted in an earlier chapter, conformation should always be the first consideration when buying any goat. Look for strong legs and feet, a well-formed mammary system for raising kids, and a good set of teeth and jaw for maintaining proper nutrition.

After considering conformation, next look at the fleece. It should cover as much of the body as possible. If you are a knitter, spinner, or crafter already using fiber, you have an advantage in deciding what you want in your herd. If you're a newcomer to fiber animals, learn as much as you can before making the plunge. It costs as much to feed a poor-quality animal as a good one.

REGISTERING FIBER GOATS

Purebred Angora goats are registered with the American Angora Goat Breeders' Association (AAGBA), established in 1900. Purebreds must be white and have purebred parents. All Angora goats need two forms of identification, whether tattoos, ear tags, or ear notching. Ear notching entails placing notches and holes into specific parts of each ear as a code.

In 1999, another registry opened, the Colored Angora Goat Breeder's Association (CAGBA). If the animal has registered parents in either CAGBA or AAGBA on both sides of the pedigree, the breeder may certify that the animal is free of disqualifying traits and register the goat without inspection. Blue Card goats are those with distinctly colored fleece, and Red Card goats have white or lightly colored fleece. Animals that do not have two registered parents must undergo inspection. Most often, this process occurs at a show or event that has inspections built into the program.

The various other fiber goats, such as Pygoras and Nigoras, have their own registries. In 2005, the Cashmere Goat Registry was established. This registry is a privately held registry, guided by a devotion to the improvement and preservation of cashmere-producing goats, over bureaucracy and politics.

CARING FOR YOUR FIBER GOATS

Most fiber goats in the United States are raised on range conditions. This method of rearing presents challenges for meeting the nutrition and safety needs of the herd. Much of the literature about fiber goats addresses this style of rearing. With small flocks, the emphasis shifts toward intensive management practices. Unfortunately, many people who keep these goats also keep sheep, and they often try to treat them as sheep.

Fiber goats are normally sheared twice a year—once in spring and once in fall. It is necessary for a fiber goat's health that they are sheared at least once a year. This allows the owner to check for external parasites as well as other health conditions that may be obscured by hair. A disadvantage of shearing only in the spring is that the goat's fiber can become severely matted over twelve months as opposed to six months.

TYPES OF FIBER

CASHMERE

- Cashmere is defined by the US Wool Products Labeling Act of 1939 and the Cashmere and Camel Hair Manufacturers Institute as "the fine (dehaired) undercoat fibers produced by a Cashmere goat (*Capra hircus laniger*)."
- The fibers have an average diameter of 19 microns, with no more than 3 percent (by weight) over 30 microns in diameter.
- It is protected by long coarse outer guard hairs that must be removed from the shorn fleece to preserve the quality of the cashmere.
- Adults produce 4 ounces of cashmere fiber per year.
- A single cashmere sweater requires the fiber of three Cashmere goats.

MOHAIR

- Mohair is the long, silky hair of the Angora goat.
- Kid fiber is 4 inches long at shearing and finer than adult fiber.
- Yearlings produce 3 to 5 pounds of mohair.
- Adults produce 8 to 16 pounds of mohair a year.

CASHGORA

- Cashgora is the hair of large-breed (Cashmere) and small-breed (Pygora and Nigora) fiber crosses.
- It has limited uses; it's typically not as fine as mohair or cashmere.
- Animals fall into three types:
  - Type A: Hair is similar to mohair and smooth and cool to the touch; requires shearing one or two times a year.
  - Type B: Hair is airy, light, fluffy, and soft and warm to the touch; harvested once a year by combing, plucking, or shearing.
  - Type C: Hair is creamy, suedelike, and warm to the touch; shows no luster; often commercially acceptable as cashmere in type; and harvested once a year by combing, plucking, or shearing.

KEMP

- Kemp is the long, straight, brittle, hollow hair that can show up on the thighs and backbone of any fiber goat.
- The fiber breaks easily and does not take dye well.
- The presence of kemp is a criterion for culling a goat from a fiber herd.

Fiber-goat owners who have colored and white goats sometimes separate the white goats from their colored herd mates a few weeks before shearing to keep from cross-contaminating the white fiber with dark. Pure white cashmere and pure white mohair have a higher commercial value than their colored counterparts. White fiber accepts and distributes dye more evenly and produces more consistent results in finished garments. This same care may not be necessary for those who provide product to the crafting market where individual character and "one-of-a-kind" handmade products serve a different market.

Goats are more prone to chilling from cold or damp weather than sheep. Angora goats with 2 inches of hair growth can tolerate subzero temperatures. However, if you expose a herd of freshly sheared Angoras to sudden temperature drops or increased wind and humidity, the animals may become ill or even die. Owners can put coats onto newly shorn goats to protect them from drafts and cold. In an emergency, a sweatshirt or jacket can serve to prevent chilling.

Cashmere goats are normally hardier. In fact, the best-quality cashmere is said to come from goats living under rough conditions high up cold mountains. The same goats brought inside and pampered would no longer produce the same quality of fiber.

HARVESTING YOUR FIBER

Cashmere

The cashmere undercoat grows throughout the fall and winter. The slight sheen on growing fleece is called "life." In the spring, as the daylight hours lengthen and goats approach kidding, the hair naturally begins to shed. The finest fibers are shed first, and the fleece becomes a dull matte in appearance.

Large commercial fiber operations tend to shear their Cashmere goats. This is done once a year in early spring, before the cashmere undercoat begins to shed. The fine cashmere fibers are then mechanically separated from the stiff guard hairs.

In small herds, the fiber is typically harvested by combing. Cashmere combs have long, sharpened tines with a bale that moves back as the comb becomes full of cashmere. A dog-grooming rake works well too. This traditional method mimics natural shedding by leaving hair on the animal and protecting against weather. Selective breeding for true cashmere and proper timing ensure that you can comb out the cashmere in one or two combings. If the guard hairs are longer than three times the length of the cashmere undercoat, trim them before combing.

Mohair comes only from Angora goats, whereas cashmere is the soft undercoat produced by almost all the other goat breeds. *Shutterstock*

Mohair

Angora goats need to be sheared twice a year, usually before breeding and before kidding. Mohair grows about ¾ inch per month on an adult goat and should be 4 to 6 inches long at shearing.

Shearing

Some breeders, especially those with many goats, will hire a professional shearer to harvest their product. When hiring a shearer, check with other breeders for recommendations. Some of these breeders are very busy during shearing season, so it is best to line up this service well in advance. Discuss the breeder's experience shearing goats. It is OK to ask the shearer to take enough time to handle the goat gently and get the best-quality fleece. A rushed clipping job can result in excess nicks and cuts on the animal as well as second cuts—shorter sections of hair that have lower value.

Breeders with smaller flocks can perform their own shearing. The same clippers mentioned in the section on showing goats can be used on fiber goats. The clipper should be fitted with a twenty-tooth mohair comb. Use well-sharpened blades.

It really isn't necessary to bathe or brush the goat before shearing. Pick out large pieces of visible junk or use a blower on the goat to reduce some vegetable matter. A stanchion or grooming stand makes shearing more comfortable for you and your goat. Take long, unbroken swaths of hair as much as possible.

Some breeders like to leave a cape—an unshorn strip of hair, 6 to 8 inches wide—along a goat's spine. A cape protects a goat from becoming chilled following shearing. Leaving the cape results in loss of some fiber. However, a cape can be a good compromise when shelter isn't available to range goats and bad weather is predicted.

Keep goats dry for twenty-four hours before shearing.

- Shear goats in order—youngest to oldest—as the coarse fiber of older animals shouldn't mix with high-quality fiber from young stock.
- Keep a clean shearing area, either concrete or plywood, and sweep between animals.
- Shear calmly and take your time to avoid second cuts (short cuts that downgrade the value of the fleece).

An Angora kid has soft, clean white hair. The cleaner you keep the fleece, the better price you will get for your fiber. *Shutterstock*

Cashmere is a highly sought product. *Shutterstock*

- Sort stained or dirty hair out of the fleece before storing.

Contamination

The cleanliness of fleece is important. There are three kinds of contamination:

- **NATURAL**: Contaminants created by the goat itself, such as black or colored fiber, yolk (a combination of sweat and grease), urine, and dung
- **ACQUIRED**: Anything picked up from the environment, such as feed, vegetable matter, poly fibers from twine, even cigarette butts
- **APPLIED**: Products intentionally applied to the goat, such as paint from brands, topical wormers, and fly spray

To lessen fleece contamination, hold the goat a minimum of four hours without feed or water in a clean, dry pen before shearing. To shear a goat, start with a clean floor or put down a piece of plywood to catch the fleece. Sweep the shearing board clean after each shearing. Cashmere goats are sheared standing up with their head in a stanchion or head bale. Cut close to the skin the first time, as second cuts result in uneven fibers that lower the value of the fleece.

Each goat's fleece is placed in its own bag. Using a permanent marker, mark the bag:

- Grower's name
- Fleece type (kid, yearling, young adult, adult, buck)
- Goat's name and identifying number
- Date
- Clip season (spring or fall)
- Problems (burrs, long fleece, short fleece)
- Fleece quality

PROCESSING FIBER

A harvested fleece can be sold "as is" or sent out to be processed. Fiber processing mills are listed online. Some mills are quite large while others are described as "mini-mills." The limited number of facilities often mean waiting lists with possibly months before your fleece can be turned into the desired finished form. Available products from these mills range from washing and batting through actual felt or yarn. Get to know local fiber-goat owners and join online discussions to learn which sources are used in your area.

Just as a dairy goat owner may get milking goats to satisfy an interest in raising their own milk, many a fiber-goat owner has purchased their goats to get closer to the source of the fiber they love. My friend Lynn, who introduced me to my first goat class at the University of Minnesota Extension, was a member of a local spinning and weaving group. Hand processing a fleece is a large amount of work. However, it can also be very relaxing and rewarding.

Cleaning and Washing the Fleece

A raw fleece smells of goat and carries not just scent, but physical reminders of its environment. The curly coat can be very clean or extremely dirty depending on whether the animal has had routine grooming and a clean stall or been out clearing brush and burdock.

Most goats have a coat that falls somewhere between those two extremes. A dirty fleece can weigh as much as 40 to 50 percent more than the cleaned product.

The greasy coating on mohair is called "yolk." Yolk is sometimes a light coating or can be a thick, waxy coating on the hair. It functions as protection from the elements to keep the fiber soft and lustrous. This grease can be hard to wash out. Kid mohair has little or no yolk, so it is easier to wash. Older animals will have more of this protective coating. Mohair with a thicker yolk is more difficult to wash, but the soft, luxurious fiber is worth the effort.

To begin cleaning a fleece, physically shake out debris and pick out stuck pieces. The loose pieces of material are referred to as VM or vegetable matter. Vegetable matter will not wash out of a fleece. However, VM will fall out of mohair fleece after drying because mohair doesn't have scales for VM to cling to, unlike wool. Then the fleece should be washed to remove dirt, natural oils, and grease as well as other impurities. Washing can be done by hand or machine.

Using a washing machine is tricky for cleaning fleece. The yolk can foul your machine. Since mohair should not be agitated, the washing machine needs to be started and stopped at specific points and water temperature should not be allowed to change rapidly. Various websites give instructions for the use of a washing machine to clean wool. Washing by hand allows better control of the process.

The easiest way to wash raw mohair is to separate your fleece into smaller bundles. Separating the fleece into sections of equal length also makes spinning easier. Put the mohair into mesh bags—the ones sold for washing delicates are a good option, or a mesh onion or potato bag. Be sure to pack it loosely to allow free flow of water and soap. It is recommended to wash small amounts (5 to 6 pounds or less) of mohair at a time. Some instructions recommend soaking the fleece overnight in a cool water bath, which opens the fiber as well as loosening the yolk. This really isn't necessary if the fleece isn't matted with grease.

Like all processes, it is helpful to collect your tools for this project. Here is what you need:

- Wash container or sink
- pH test strips—available from pool or brewing supply stores
- Detergent—Dawn dish soap or a product made specifically for cleaning wool
- Rubber gloves
- Mesh bags
- Baking soda
- Drying rack

Here is the cleaning process:

- Rinse the fleece with running water to remove surface dirt—this can be done outside with a garden hose,
- Run hot water—145 to 160 degrees Fahrenheit—into your wash container. If you can put your bare hands into this water, it isn't hot enough—you will be using gloves.
- Test the pH of your water. Neutral pH is 7.0, so the water may need to be adjusted. For pH below 8, add small amounts of baking soda and mix well until the pH is 8 or below. This helps on the first rinse.
- Mix soap into the water.
- Submerge your mohair, making sure it is completely saturated and floats freely.
- Soak for 10 to 15 minutes, gently turning and dunking the bags. Be careful: mohair doesn't felt as easily as wool, but it can be damaged if handled too roughly.
- Check a lock of fleece—if it is gummy, the grease hasn't soaked free and you may need to keep soaking up to 45 minutes.
- Maintain a water temperature over 145 degrees Fahrenheit so the yolk doesn't readhere to the fiber—it is difficult to remove after this happens.
- Drain the water and squeeze excess fluid from the fleece.

- Refill the container with hot water plus about half the original amount of detergent.
- Soak another 10 to 15 minutes.
- Check a lock of fleece again—rinse the lock with hot water to see if your fiber looks clean. If the mohair or water looks milky, there is more grease present.
- Drain, squeeze, and repeat the rinse as necessary until the water runs clear, using progressively cooler water.
- Check the pH of the final rinse—for a pH higher than 6.0, add a small amount of vinegar to lower the pH and soak for a final fifteen minutes.
- Remove the bags and press them gently to remove excess water.
- Remove the fleece from the bags and spread to dry on a towel or laundry rack like you are drying a woolen garment.
- Flip the fleece periodically to dry it evenly.
- Leave the fleece in the sun or use a fan to speed up drying.

Carding

Once you have a clean, dry fleece, the fiber needs to be carded. Traditionally, fleece is prepared for spinning using hand paddle brushes to separate and straighten the fibers. The resulting batt of fluffed up mohair makes spinning easier. Hand-carders are wooden paddles with wire faces. These carding combs come in two sizes, coarse or fine. The coarse paddles are the size primarily used to card mohair.

Using your Mohair

Processing your fleece is a starting point for many adventures. The products to be created are limited only by your imagination. If you are new to fiber arts, consider checking out online video instructions for using an inexpensive drop spindle. The drop spindle or more expensive spinning wheel allows a goat owner to turn their fleece into yarn for knitting, crocheting, or crafts. The art of felting also starts with natural fibers.

FIBER MARKETS

Just like goat milk, goat fiber is a specialty commodity that has various outlets. Your area may have a local cooperative that fiber producers can join to market their bulk fiber. Some producers sell fleeces directly to handspinners for up to $12 per ounce; others process their own fiber and sell it as yarn or other products. Commercial facilities will process fiber for a fee, after which the farmer sells the cleaned mohair or cashmere.

Commercial herds often sell the fleece as it has been sheared from the goat. The Texas wool warehouses and other commercial buyers that handle commercial goat fiber will not take colored fleece. Producers are gaining profits from added-value processing. Small producers or those with colored herds market differently, often to crafters. Washing and carding the fiber on farm allows for an additional $4 or more per pound when sold to spinners. Spun into yarn and sold natural or already dyed, the same pound of fiber can be sold for $80 or more.

Secondary Markets

Don't forget that when you raise a fiber herd, you also have a secondary meat market. Excess males and goats that do not produce quality fiber are a fiber industry byproduct. The Angora goat herds of the Southwest are a major source of meat goats. Many agricultural census tallies have Cashmere goats counted in the meat-goat category.

ENJOYING YOUR PET GOAT

Goats are smart and personable. They pair the friendliness of dogs with the intelligence of cats. Inquisitive by nature, goats explore new surroundings, test fences and gates, try tasting everything, and generally poke their noses into whatever is within reach.

These qualities make goats wonderful pets—or real problem animals, depending on their upbringing. Just as an undisciplined dog causes problems for the owner, a goat becomes a nuisance when untrained.

Pet goats such as this Nigerian kid can be wonderful companion animals. *Jen Brown, Cutter Farms*

RULE 1: TEACH BASIC COMMANDS

A goat can learn simple commands just as a dog can. Choose simple words and use them consistently to reinforce behavior. When training, accompany the commands with direction from your hand or a squirt from a water bottle. Some form of physical contact should be used at first. Chasing the goat is counterproductive and may teach it a new game: running away! Some basic, useful commands include:

- "No": Stop what you are doing.
- "Stop" or "Back": Don't push and don't rush the gate.
- "Come": Move toward me.
- "Hup," "Hey-up," or "Up": Jump into a trailer or onto a milk stand.

RULE 2: NO PUSHING

Whether or not your goat has horns, allowing it to play by pushing with its head is a bad idea. Goats settle disputes between each other by butting heads. When you push back at the head, you teach the kid that this is an acceptable way to play with humans.

Never invite head pushing or butting. Do not grab the goat by its horns. When the goat lowers its head, hold out your hand and issue the command "No!" or "Back!" Gently redirect the goat by placing your hand on the side of its face. If the goat continues to butt, accompany your command with a squirt from a water bottle or a sharp tap on the end of the nose. Never hit the goat, or it will learn to fear you.

RULE 3: NO JUMPING

Pet goats should not jump on people. To address this behavior, gently grab the goat by the shoulders and set its feet firmly on the ground. Accompany this physical contact with the command "No!"

For persistent bucks, it may be necessary to be more forceful. Try grabbing the goat when he is in midair and forcing his head and shoulders all the way to the ground. Touching his head to the ground teaches him that you are dominant.

RULE 4: CONTAIN THE GOAT

Over the years, I have heard these stories from people visiting my farm to give up their goats. Unsupervised goats running loose in the yard will get into trouble. When supervised, the goats

A well-disciplined goat is a happy goat. One basic command is "Hey-up," meaning jump up, in this case onto a hay wagon.
Jen Brown

"WE HAD A GOAT WHEN I WAS A KID AND HE ALWAYS CLIMBED ON CARS."
"ALL OF MY ROSES ARE GONE, AND THE LILACS HAVE BEEN CHEWED OFF TO THE GROUND."
"THE GOATS GOT INTO MY NEIGHBOR'S TREES AND STRIPPED BARK OFF THEIR TRUNKS. SOME OF THEM WERE SPECIAL ORNAMENTALS, SO IF THEY DIE, I HAVE TO REPLACE THEM."

can be redirected from bad behavior. Alone, goats should be confined to an enclosure.

Tethering or tying a lone goat out on a lead is not a good idea. A goat staked out alone has no protection from stray dogs or predators. He could also become tangled and hurt or scare himself.

RULE 5: KEEP A FRIEND

Goats are herd animals. A solitary goat becomes lonely without constant companionship. Usually, this means the goat owner should keep at least two goats. However, a goat will bond with almost any species of stall mate, including a horse, sheep, or cow.

Left completely alone, a goat will develop a number of negative behaviors. Some goats cry. Lonely Nubians in particular have very loud voices. A crying goat quickly becomes as annoying to your neighbors (and you) as a barking dog.

Allowed to run loose, a lone goat will hang close to the house in order to get close to the people it thinks are its herd. This means your deck or doorway may become fouled with nanny berries. Other goats will seek out other farm animals or run away in search of companionship. Some of these goats get loose on their own, often damaging fencing or housing in their desire to have friends.

Any breed can make a good caprine pet. Nubians are beautiful and popular, but every breed has something to offer. *Jen Brown, Terrapin Acres*

Goats are playful and need to be active to stay healthy. An outside pasture or yard can be made more appealing by providing distractions. *Jen Brown*

Left to their own devices, goats run wild just for fun—or tussle to decide who is higher in herd status. *Jen Brown*

THE GOATS HAVE TAUGHT ME A LOT IN THE PAST THIRTY YEARS. THEY DON'T, FOR EXAMPLE, CARE HOW I SMELL OR HOW I LOOK. THEY TRUST ME AND HAVE FAITH IN ME, AND THIS IS MORE THAN I CAN SAY FOR A LOT OF PEOPLE.

—CHES MCCARTNEY, THE GOAT MAN

Goats and horses have a natural relationship that dates back to the early days of horseracing, when small, cheap, and easy-to-transport goats were kept as stall companions to calm down high-strung racehorses.

A Pet for Your Horse

Goats make good companions for horses. Just as a goat gets lonely, a single horse pines for friendship. The phrase "to get his goat" originated from the practice of stealing the goat companion of a racehorse before the race. Any size goat can be paired with a horse. The advantage to using a full-size breed is that the goat is closer in size to the horse.

Introduce the goat and horse gradually by penning them in adjoining stalls. Use caution during the first few weeks, while the animals are adjusting to each other. Be aware of possible problems and be ready to separate the animals if necessary. Horns on a goat can be dangerous for the horse. Some goats like to chew on the horse's tail. An aggressive horse may kick the goat or pick it up and throw it.

Fencing for horses is frequently not goat-proof. Adding lower electric strands or making a smaller fenced area for the goat may be necessary.

FUN STUFF

One of the joys of owning goats is their playful manner. Goats love attention, so the tricks and games you can play are limited only by your imagination. Some of these games have become formal competitions at fairs. Agility and costume classes at shows are fun for goat exhibitors and spectators alike. Elaborate or simple playgrounds provide exercise for the goat and keep the owners amused.

Agility Obstacles

These are some of the obstacles you might see at an exhibit.

- **BALANCE BEAM:** A plank suspended a few inches or a foot off the ground that the goat must walk across without stepping off
- **BRIDGE:** A small bridge that the goat must cross
- **JUMP:** A raised bar, crossed logs, or other material that the goat must leap over
- **HANGING TIRE:** A tire suspended in the air that the goat must jump through.

- **SLALOM**: A zigzag course made of poles that the goat must weave through without skipping or knocking any over
- **TEETER-TOTTER**: A plank balanced on a central pivot so that the plank tips as the goat crosses
- **TENT**: A tent the goat and owner must enter
- **TIRES**: A course made of tractor or truck tires, placed flat on the ground, that the goat must step into and out of as it crosses
- **TUNNEL**: A vinyl or hoop tunnel that the goat must pass through
- **WATER HAZARD**: A child's wading pool, stream, or large puddle the goat must step into

It's best to have at least two pet goats for the sake of the animals. A solitary goat is often an unhappy, noisy, troublesome goat. *Shutterstock*

PACK AND HARNESS GOATS

Goats can pull a cart or carry a pack. Itinerant traveler Ches McCartney, otherwise known as the Goat Man, famously wandered the United States with a wagon pulled by a team of goats from the 1930s to the 1960s. Some travel companies use pack goats for expeditions. Goats are perfect for moving about rough land where other animals find impossible to maneuver.

Size is the main factor limiting the pulling or carrying capacity of a goat. A well-conditioned pack goat can carry up to 25 percent of its weight. The same animal should not be asked to pull a cart more than one-and-a-half times its weight.

Training a goat to work is neither easy nor quick. The effort pays off with time and patience. Goats love attention and company, so those trained to pull or carry will soon enjoy the activity as much as you do.

SHOWING YOUR GOATS

I once believed that showing animals was all about vanity. Admittedly, when you have a particularly nice animal that seems unbeatable, it is hard to remain humble. But the best reason to show has nothing to do with pretty ribbons, trophies, or champion legs. It's to compare your animal with other goats. Getting other people's opinions about the goats in your barn and sharing war stories with other breeders is an education in and of itself.

My first two Nubians, Celeste and Menolly, were meant to be milk and hobby-farm animals. I had no interest in showing goats. Linda Libra convinced me to bring the "girls" up for a day at the East Central Dairy Goat Association Show. I labored to clip just enough hair (but not too much!) so that the kids looked neat.

The weather was cold and rainy. Goats and people seemed to be everywhere. Talk was indecipherable. My kids looked shaggy compared to the other, perfectly trimmed animals. Several breeders kindly commented on how smart I was to leave them "fuzzy" so they wouldn't catch a chill. The whole day was confusing but fun. When Celeste took first-place senior kid, I was hooked on the world of goat shows.

WETHERS AS PETS

The most common pet goat is a castrated male known as a wether. This popularity is mostly by process of elimination. A breeding buck smells bad and has all sorts of unappealing behaviors (such as peeing on his face!). A doe needs to be milked and comes into heat every fall. Breedable animals tend to be more expensive than neutered males.

The best place to buy a wether is from a reputable breeder. A breeder may charge as much as $200 or more, but this typically gets you a neutered, disbudded, vaccinated animal plus after-sale support. You can also buy wethers from auctions or private parties, such as Craigslist. Buyer beware. Free or cheap goats can be either a tremendous bargain or a heartbreaking mistake.

The wether has special diet considerations. Unlike a buck or a doe, a wether does not expend energy on mating or milking. An overfed wether quickly becomes fat. Limit his grain and treats. A wether is also prone to urinary calculi, a potentially fatal condition. To help prevent it, supplement his feed with ammonium chloride—and keep his weight down!

Empty buckets sometimes get used as toys, but goats don't like being in the dark and will become spooked if the bucket gets stuck. *Jen Brown*

The world of showing goats offers many possibilities. Shows run the gamut from "fuzzy trailer shows"—where the goats are unclipped, brushed, and shown right out of the owner's trailer—to public events where the goat and handler are at their cleanest and best. County and state fairs hold shows that pay modest premiums for animals shown in either open class or some of the youth-centered programs, such as 4-H and FFA. Breed associations sponsor shows. Most of these organizations hold at least one national show each year. In the case of pet goats, sometimes fairs have fun classes in which the goat and owner show together in a costume contest, obstacle course, or cart exhibition.

FINDING SHOWS

Contact the breed registry for whichever type of goat you want to show. Check fair premium books or club websites. Become a member of a local goat club to get news of upcoming shows and events.

Once you are integrated into the caprine network, finding shows is fairly simple. Unfortunately, in some areas, there are no nearby shows. Then you need to decide if you want to travel. For more and more states, there is an added cost to crossing state lines with goats. State regulations may require travel and health documents, which are discussed in chapter 2.

A pack goat can pull a cart for exhibition or work around the farm. Hoppy the cart goat was always popular at the Washington County Fair in Minnesota. *Carol Amundson, Hoppy Land*

TO CLIP OR NOT TO CLIP?

Fitting requirements vary depending upon the show, the type of goat being shown, and even the weather. In some areas, the goat is lightly trimmed—some find a ⅜-inch clip preferable—and sometimes a meat animal is "slick sheared," or clipped like a dairy goat. Some breeds of goats should not be clipped before the show. In the case of fleece shows, the goat may not go at all, only her wool. Miniature goats and meat goats are often shown unclipped. If Guernsey breeders in the United States gain approval for their breed to be recognized and shown as a dairy breed, they are considering the recommendation to only clip the udder, leaving the long silky coat as part of the show clip.

DAIRY CLIP

A dairy goat probably does not have one bit of skin left untouched by the time she is fully prepared for a show. To begin grooming your goat, tie her to a fence or lock her into a milk stand. Bathe her first with a livestock shampoo that has antifungal properties. Use a scrub brush to work out loose hairs, dirt, and debris. Rinse well. Follow with a skin and coat conditioner. If the weather is cool or windy, cover the goat with a coat or blow her dry so that she won't catch a chill.

The best trimmer for the body is a large livestock clipper. Blade marks can be seen on the body of the goat more or less readily depending on her color. Black or dark goats should be clipped a few days to a week before the show to allow time for the clipper marks to disappear. White goats should be clipped as close to the day of the show as possible.

Clipper Care

Your goat's clip job is only as good as your equipment. Sharp blades and a well-lubricated clipper make the job easier. Clippers get hot.

A child interacts with goats on display. *Shutterstock*

Each exhibitor brings feeders and other gear to make the goats feel comfortable in their temporary home. *Carol Amundson*

GOAT SHOW PACKING CHECKLIST

LIVESTOCK AND GEAR

Barn lime
Bedding straw or shavings
Broom
Clippers
Fasteners for feeders and display
Feed
First-aid kit (caprine)
Goats (of course!)
Grooming supplies
Hay
Hay and grain feeders
Hoof shears or knife
Milk stand
Milking equipment
Pen sanitizer
Rake
Pasteurizer
Shampoo and conditioner
Tools

SHOW SUPPLIES

Blades
Blade wash/clipper lube
Brush for cleaning blades
Large clippers
Small clippers

PERSONAL GEAR

Black shoes, white show clothes
Buckets
Coffee pot and supplies
Cot
Food/snacks
First-aid kit (human)
Pillow
Sleeping bag
Tent
Toiletries
Street clothes

Be prepared to switch clippers or let the animal rest as the blades cool down. Use a small brush to clean hair from air vents and blades. Use a clipper lubricant, such as Kool-Lube, frequently. Kerosene and alcohol combined in a coffee can makes a good cleaner and lube. Dip the clippers periodically while running to clean the blades. Lightweight sewing machine oil can be dripped onto oil ports. Before putting the clippers away, clean thoroughly.

DAIRY-CLIP EQUIPMENT

- Blades (#10 and #30)
- Blade wash/clipper lube
- Brush for cleaning blades
- Large clippers
- Small clippers

The Body

Use the wide clippers on the larger parts of the body. Be prepared to change the angle of the clip on parts of the body where the hair grows in another direction. Some goats are ticklish, especially on their belly, under the legs, or around the feet. A firm touch helps.

1. Set the clipper at the lowest setting that doesn't pull the goat's hair.
2. Start along the goat's back on either side of the spine.
3. Shave from tail to head against the hair growth in long, smooth strokes that go down across the body.
4. Stretch the skin taut over the hip bones to get into the pocket in front of the hips.
5. Hold the goat's front legs up to shear underneath the chest and along the stomach. Hair in this area is easy to miss.

The Feet and Legs

The legs should receive a lot of attention, especially on young stock. More points are awarded for feet and legs in judging dry stock than in judging does in milk. This is the time to switch to a smaller grooming clipper. (I use an

Of course, sometimes the goats just get tired of all the attention. *Carol Amundson*

Nubians wait their turn in the show ring. *Shutterstock*

Oster A-5.) With the 10 blade, the hair should be clipped to show off the feet and legs.

1. Shave the legs from the bottom up, against the hair growth.
2. To get at the pockets and angles of the hind leg, push the skin out at the hollow points.
3. Remember to shave under the legs. You may have to hold the clippers upside down and look hard to confirm that this area is clean.
4. Pay special attention to the pastern (the back of the leg above the hoof) and between the toes.

The Head

Some goats are very touchy and don't like having their head clipped, while others love it. One of my older show goats used to turn her head so I could "scratch" her just right. It may be helpful to take the goat out of the stanchion and straddle her back.

1. With the small clippers, shave as much off the face and ears as you can.
2. Use long, smooth strokes from the base of the neck up to the chin.
3. Aim for a smooth cut across the neckline, blending down to the chest and body.
4. Trim the poll (the knobby top of the head).
5. Pay special attention to the ears on goats that have ear tattoos. Clean the hair from inside the ear and make sure the tattoo is legible.

The Tail

1. Trim the tailhead and around the base of the tail.
2. Holding the tail by the tip, shave toward the tailhead, leaving about 2 inches of hair on the tip (called a puff).
3. Trim the end of the bush straight across. (Some breeders leave kid tails ending in a little pom-pom instead.).
4. On LaManchas, this is a good time to check the tail tattoo.

The Udder

For the udder and escutcheon, the hair needs to be neat and very short. Some breeders use surgical blades or disposable razors. It is easier to shave a tight, full udder. Clip the udder the evening before the show while it is full of milk, and do any needed touchups in the morning. The udder is worth points on the scorecard. The udder area and escutcheon on dry stock are also important.

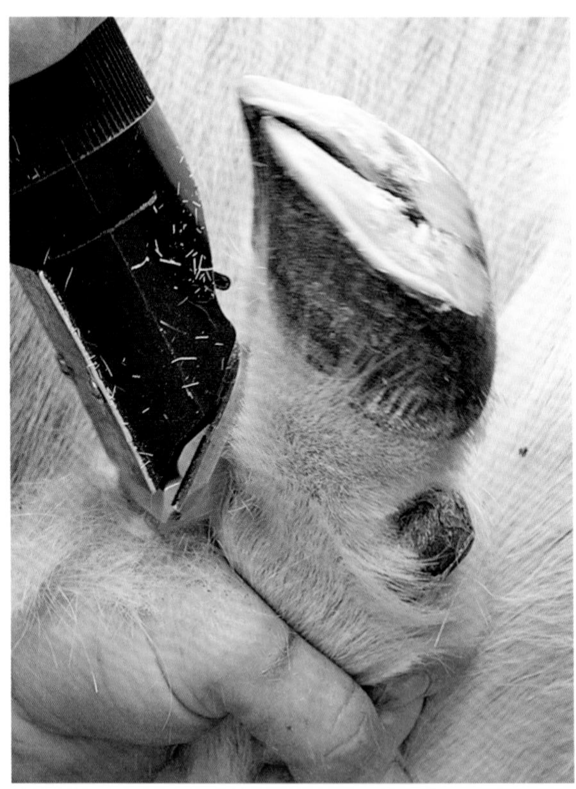

Above and right: Clipping for the show is usually done beforehand at home, but having the clipper and lube available at the fair helps for last minute trimming of pasterns or udders. *Jen Brown*

Above and right: Some local fairs have a costume show, a fun category judged on originality, comfort, and ease of movement for the animals. Normally the goat and exhibitor must both be in costume. *Carol Amundson*

Watching shows helps you learn show techniques and etiquette. Goats are moved at the judge's direction. As the goats are walked around the ring or stopped for inspection by the judge, exhibitors should always be aware of how the goat appears to the judge. *Carol Amundson*

1. For the first trim, use the body clippers with the 10 blade to remove long hair.
2. Use a 30 or higher blade to shave the entire udder, including under the goat's legs and between the two halves of the udder.
3. Shave the escutcheon about ½ inch above the udder attachment area in a curved shape. This should include the attachment areas around the back of the legs to make the udder appear wider and fuller from the rear.
4. Trim the fore udder to emphasize the attachment and the milk veins.
5. Some breeders apply a cream or skin softener. Skin So Soft helps keep bugs away. Be certain not to use greasy cream, which will attract dirt and rub off on the hand when the judge checks the udder.
6. Recheck the udder grooming the morning before the show to catch any hairs that were missed and to touch up the attachment area. Take care that the udder isn't overfilled and relieve excess fullness if necessary.
7. Use teat tape to prevent leakage on does that have loose orifices; however, tape should be removed before the show. Do not use glue or other substances to block the teat.

SHOWMANSHIP TIPS

- Train your goat to lead and stand properly before the show.
- Walk your goat at a steady pace. Don't run or move too quickly.
- Watch the judge. Always know where the judge is and what is expected of you.
- When possible, walk your goat into the showing area before the show so it can become familiar with the surroundings.
- Hold your goat's head high with the collar held under the chin. A choke-chain collar is most commonly used in the show ring. This is not a safe collar to leave on an animal, so be certain to remove it after the show.
- Keep the goat between you and the judge. This is known as the "peanut butter sandwich." The goat is the peanut butter, and you and the judge are the bread.
- Move in front of the goat (never behind) when turning.
- Be aware of your goat and set it square when standing still. Don't overhandle your animal. Once it is set, avoid distracting the judge from looking at your goat with a lot of movement.
- Don't crowd the goat ahead of you. Keep a few feet between your animal and the next one so the judge can look at each goat properly. This buffer also prevents tail biting and other negative contact between unfamiliar goats.
- Know your goat. The judge may ask for the animal's age or birth date, kidding date, or number of kiddings. For showmanship class, be prepared to answer more technical questions about management, parts of the goat, or how you feed, vaccinate, and worm your project animal.

Agility obstacles are another fun category at goat shows. Slalom poles and a teeter-totter are but a few of the challenges placed before the goats and their handlers. They are judged by time and skill. *Carol Amundson*

Anna Thompson-Hajdik, of Legendairy Goats, milks out her Alpine doe following competition at the Minnesota State Fair. Milking at the fair makes for great public relations. *Carol Amundson*

FINISHING

Just before your goat goes into the ring, touch up any manure spots or other dirty areas. Wet wipes or a damp cloth are useful for last minute cleanups.

1. Check and wipe under the tail and around the anus.
2. Brush off loose hairs and bedding.
3. A light spray with Show Sheen or other conditioning spray keeps the goat looking her best.
4. A blanket or coat on your goat will help keep it clean after final grooming.

AFTER THE SHOW

You may want to use antifungal soap on your goats after the show to prevent ringworm, which could have been picked up at the show. Isolate and treat any goats that exhibit diarrhea or respiratory illness.

JUDGES, JUDGING, AND KEEPING IT ALL IN PERSPECTIVE

Shows are subjective. Each registry tries to make showing more objective through its scorecards and judges' training, but it isn't easy to select the best goat from a lineup of beautiful animals. The doe that stood first yesterday may not place as high today and vice versa. One judge's opinion is just that, an opinion, no matter how many years he has been judging or what scorecard he is using. Before you start to show your goats, I strongly recommend going to shows in your area. Talk to exhibitors. Watch the judges and listen to their reasons.

Shows are fun and exciting. They can also be exhausting, frustrating, and downright discouraging. Each show is a new experience. One of my friends, who has been showing goats since her son was in 4-H (he now judges and has his own children), has one of the best show perspectives I have ever encountered. Her philosophy is "We show what we've got." She continually works toward breeding the best goats she can while raising animals she likes to have around.

MARKETING YOUR GOATS AND THEIR PRODUCTS

Live animals, milk, meat, hair, and hide can all be marketed, depending on your supply and local regulations. Here is an overview of some sales venues and how to approach them.

AUCTION OR SALES BARN

The sales barn or local livestock market auction is a way to sell animals with minimal fuss. In some parts of the country, these markets are closing and consolidating, so distance will be a limiting factor. Here in Minnesota, a number of the livestock facilities have shut down. Even the South Saint Paul Stockyards, after 122 years in business, closed around the time the first edition of this book was being printed.

However, if you are close to such a facility, you simply drop the animals off and receive payment at the end of the auction. Live market auctions sell goats by the pound or by the head. In the case of stockyards with buyers on premises, you may even get their market price paid to you directly at the time of delivery without having to wait to sell the animal first. Rarely do you receive top price this way, but it is a good way to dispose of culled animals.

Goats and their products can be marketed through many venues. *Shutterstock*

Specialty auction sales are coordinated around exotic animals or specific types of goats such as meat or fiber. These auctions sometimes offer animals with registration papers. Depending on the auction setup, the animals are of higher quality or give more information about the goats for sale than a sales barn-type sale. Sometimes, local goat clubs may hold fund-raising auctions. Breeders will put up some very nice animals for consideration at these sales with some or all of the proceeds going to the club.

DIRECT MARKETING

Direct marketing to individuals is probably the most common method of selling live goats. Goats and their products are specialty items with a market best reached through targeted advertising rather than mass marketing.

Signs and Bulletin Board Notices

An attractive farm sign identifies your location and attracts drive-by business. Most agriculturally zoned properties allow sign advertising on their property without community approval.

Bulletin boards at the feed mill, grocery store, or pet store are other places to post sales bills. If you have a product to market beyond live animals, you are limited only by your imagination. The grocery store, the health food market, and ethnic gathering spots are all worth trying.

Print Advertising

Print ads are paid by the word, line, or page layout (such as a quarter of a page). Be sure to track which ads bring in buyers before spending too many advertising dollars in this area. At minimum, your ad should contain the basics of what you are selling (goats, yarn, meat, and so on.) and contact information. I include my farm name, phone number, and address or general location. If you have a website, this is a great opportunity to direct people to the site.

Local newspapers will get the ad to buyers right in your area. Most papers have an online presence and will list your ad in both places. Specialty farm newspapers target your market even more precisely.

Magazines can be useful for a general monthly farm ad. Don't limit yourself to just the industry magazines. Goats are popular with back-to-the-landers, homesteaders, and anyone interested in organic and sustainable products.

Local goat associations often have a club newsletter. Here in Minnesota, I enjoy the Gopher Goat Gossip to find out what is happening on the state dairy-goat scene. It is also a source of inexpensive advertising.

Dairy products (top), goat meat (middle), and fine fiber (bottom) are some of the products a farmer can obtain from goats. *Shutterstock*

Internet Marketing

Marketing online can be as inexpensive or as costly as you want, depending on your approach and internet skills. The ability to link directly to your target market is a big advantage.

A well-designed personal website advertising your farm allows you to reach buyers with as much or as little information as you have the time and inclination to provide. You can share your farm's history, photos of your farm and animals, pedigrees, your breeding philosophy, and more.

Cybergoat, the Goat Connection, GoatWeb, and GoatFinder are just a few of the goat-specific websites that have places to list your farm or advertise products and animals. Khimaira Web Hosting grew out of Linda Campbell's Khimaira Farm business and currently serves as host for many goat and goat-related websites.

Craigslist is an online classified ad service that offers no-cost advertising. It's a great place to sell caprine products, such as goat-milk soap, hide, hair, and even leftover equipment. While pet sales are prohibited by Craigslist rules, it is legal to list livestock for sale.

Facebook prohibits sales of animals. However, there are many groups related to farm animals in general, local farm communities, and so on. On some of these sites, animal sales get by with "adoption fees" for animals or by listing pricing information in the comments of the post rather than the post itself. Even if you aren't selling through social media, you can create a farm page or participate in discussions or forums online, so your name is known.

Farm Stand

A farm stand—at a farmer's market, at a craft show, or in your yard—reaches people looking to buy direct from the producer. You can sell raw fiber and hides or processed items, such as goat-milk soap and cashmere scarves. It is not common to sell livestock at a stand, and local laws regulate the sale of products for human consumption, including goat milk and meat.

Be sure to know local regulations before selling food or cosmetic-related items. One example is raw milk. In some states, sales of raw milk are strictly prohibited. In Minnesota, you can sell small quantities of milk direct to the consumer, but you may not advertise this. One way around that is a sign that says, "Dairy Goats," a message that implies milk without directly advertising it.

Products made with milk, such as cheese, fudge, or yogurt, may fall under labeling or home-kitchen state laws in addition to milk production laws. If you are selling a product like soap, be sure you label according to local requirements. We live in a litigious society and there are many rules and regulations surrounding homemade products. These rules shouldn't discourage you from selling. However, do your homework before putting your product out to the public.

A pen containing pet goats is always a hit with visitors to our market garden. *Carol Amundson*

ASSOCIATIONS AND AGRICULTURE COMMUNITY LINKS

Your state department of agriculture may have marketing programs that provide low-cost or free marketing for producers. Contact your local county extension office, listed in the telephone book under *Agricultural Services* in the county listings, or your state department of agriculture to investigate whether your state offers sale opportunities.

These are the same organizations that can provide aid if you want to sell products that require licensing or special labeling.

Shows

Have a presence at livestock shows and keep track of your goats' wins and placings. An attractive display at these events is a useful marketing tool. This method is time-consuming and costly. It also reaps rewards for the diligent breeder who wishes to sell breeding and show stock. Goat shows allow breeders to talk to one another, see animals at their best, and hear the opinions of trained judges.

Sales List

Compile a list of the goats you have for sale, including each animal's age, sex, price, and other important details. Distribute your sales list at fairs and shows and be sure to collect the sales lists of other breeders. The breeder with whom you exchange lists may very well be a customer for outcross animals at a later date. If you don't have any goats for sale during a particular show, collect names and addresses so you can forward information as animals become available. Remember, the goat community is relatively small, so keep cordial relations with as many breeders as possible.

Word-of-mouth exchanges can be very helpful—or they can ruin your reputation. I

THEY WHO DON'T KEEP GOATS AND YET SELL KIDS, WHERE DO THEY GET THEM?

—SPANISH PROVERB

An enterprising youth group at the fair tells people about goats and lets them try their hand at milking. *Carol Amundson*

A kid pickup is a valuable community effort to sell excess dairy or fiber kids for meat. *Carol Amundson*

regularly get referrals from other goat breeders who either have no animals available or don't raise the same breeds I do. In the same way, I pass buyers on to others.

Kid Pickup or Marketing Pool

Farmers sometimes pool their goat kids and other meat goats for holiday markets. Called kid pickup or a marketing pool, this event occurs in the spring for Easter and Passover in the upper Midwest. Timing may vary in different parts of the country. When I was dairying, Daniel Considine of Sunshine Farms had contacts within the goat community. I would see a message in our club newsletter or would get a letter with dates and specifics about when he would be collecting animals. Growers meet at a predetermined location with their excess bucks, wethers, and other kids suitable for meat.

Producers can also get together a group of similar goats to sell to a volume buyer. The animals are weighed at a central pickup point. Some co-ops use this approach and negotiate prices based on the higher volume of goats.

Cooperatives

Goat cooperatives have been formed for selling caprine items, such as milk, cheese, meat, and fiber. If you raise a small herd of goats, it can be tough to keep up with market demand. Goat breeding is seasonal in nature, and an individual farm creates a relatively small amount of saleable product. Forming a cooperative helps to increase product supply while evening out some of the seasonal variation. It also allows co-op members to pool resources for advertising, processing, and shipping.

SALES ETHICS

Marketing your goats can be a hobby or a business—or just a byproduct of having more goats than you need. Whatever the case, treating your animals, your products, and your buyers as you would want to be treated isn't just ethical, it's good business. Ethical sales spur repeat business.

If you belong to a goat registry, it most likely has a code of ethics or a trade policy for selling goats. These guidelines can be useful for creating your own set of rules on your farm. Outlining your promise to your customer in a simple sales agreement can go a long way toward avoiding misunderstandings:

Disclose any problems up front. No animal is perfect. If there are defects, be straight with potential buyers. If a doe has been a disappointing milker, a poor mother, or problematic in some other way, provide the information up front. You may still make the sale. I have bought problem goats because they had other features that appealed to me.

Educate new owners. When confronted with a brand-new goat owner, expect to answer lots of questions. Tell them the basics of feeding, housing, and caring for their new livestock. This can save you time answering panicked calls later and may save the goat you are selling from injury—or death. Willingness to provide follow-up advice is the sign of a good breeder. If you just want to unload a low-quality goat, take it to the sales barn.

Focus on a "fit" and talk enough with your buyers to know what they want. You want to sell your goat or product. However, it doesn't make sense to sell an animal to anyone whose needs won't be met by the sale. Spur-of-the-moment sales are never a good idea. In the best-case scenario, the animal returns home none the worse for wear. At other times, goats or their products get a bad name, good-quality animals end up in the sales barn, and you get negative word-of-mouth.

Identify the animal clearly by making sure tattoos, ear tags, or other IDs are present and legible. Premises ID and regulations aside, it is helpful to know which goat you are selling. Tattoos can fade—or possibly were not there to begin with. Ear tags pull loose. Even microchips migrate. The goat leaving your farm should be the one you want to sell and the one the buyer wants to own. Accidents happen on the best of farms. A simple check at the time of loading prevents problems down the line.

Maintain good records and provide them to buyers when necessary. Pet-goat owners may want nothing more than the age of their goat. Buyers of breeding or show stock require more—everything from registration papers to show and production records. Have this information in an organized and easily accessed system.

This practice is especially important when selling breeding services. Provide the service memorandum at time of service. Disclose any problems with papers before the sale or breeding is finalized.

Never repeat negative gossip about another farm—and think twice about telling a negative anecdote about a firsthand experience. It is good practice to worry only about your own reputation, not that of your competition. Negative information overheard about another farm may be true—or it may be the result of a misunderstanding. Repeating stories rather than focusing on the strengths of your operation can reflect as poorly on you as on the one whose faults you expose.

Sell healthy animals. To some breeders, assuring that goats are healthy means testing for known diseases, raising kids using CAE prevention techniques, and maintaining a closed herd. To others, it simply means making certain that the goat exhibits no obvious signs of illness and is current on vaccines and hoof trimming. Spell out your health management practices to prospective buyers. Some sellers give vaccines, worming, and hoof trimming at the time of sale to show the buyer how those tasks are done. This is also tangible proof that you have done them.

Use responsible management and proper care, including housing, feed, health care, and

Above and left: Take care to follow a code of ethics when selling a goat to a buyer. *Shutterstock*

maintenance. Your goats and the products provided by them are only as good as the care they receive. Keeping up-to-date on goat management is an ongoing process.

Regardless of whether your management style is conventional, organic, or somewhere in between, it is good business to take good care of your herd.

RESOURCES

MANAGEMENT CALENDAR

This calendar recommends general management activities for goats. Owners should modify the calendar to fit their breeds, management style, and environment.

FALL

Bucks

- Clip excess belly hair.
- Examine penis and testicles for injuries and inflammation.
- Shorten or remove scurs before breeding.
- Treat as recommended by semen processor if the buck's semen will be collected.
- Turn out bucks with does at a ratio of 1:25 to 1:50.
- Use a raddle harness on bucks before pasture breeding for improved tracking of breeding dates.

Does

- Continue to flush does for two to three weeks after they are penned with a buck.
- Culture milk to pick up subclinical mastitis.
- Examine udders for signs of mastitis or injury.
- Plan dry-treatment if mastitis has been a persistent problem; treat dry does when kids are weaned or milking stops.
- Reduce grain for does that are too fat. It is easier and safer to change body condition before dry-off.

Kids

- Choose replacement does and bucks.
- Evaluate final kid crop.
- Wean kids.

Herd

- Treat any lice or skin disorders.
- Cull unsound or inferior animals.
- Determine body condition scores.
- Increase grain for two to three weeks before and after breeding.
- Give supplemental selenium and/or copper, if necessary, two to four weeks before breeding.
- Record breeding dates.
- Monitor internal parasites by testing stool samples.
- Perform any dehorning procedures after frost if season permits.
- Plan winter supplemental feeding program.
- Test herd for diseases of concern.
- Cull or isolate test-positive animals from the clean herd.
- Treat for intestinal parasites as necessary after a hard freeze or before moving to new housing.
- Trim feet and monitor for foot rot before rainy season.
- Vaccinate for tetanus and enterotoxemia (CD/T) unless your veterinarian recommends a different schedule.

Housing

- Clean and disinfect pens and barn before the start of bad weather.
- Evaluate housing and repair as necessary, paying special attention to leaky roofs and drafts.

Pastures

- Evaluate range and forage conditions.
- Move herd to fresh green pasture, if available, before pasture breeding begins.
- Soil-test pasture and supplement as needed.

WINTER

Bucks

- Remove bucks from breeding pen or pasture after does have settled.
- Supplement diet to regain body condition lost during rut.

Does

- Delouse fiber goats at shearing if necessary.
- Dry off and dry-treat udders sixty days before due date.
- Give vaccine boosters three to five weeks prior to due date.

- Gradually increase quality of hay and amount of grain fed to late-pregnancy does.
- Perform pregnancy testing; the best window for ultrasound evaluation is forty-five to sixty days post breeding.
- Shear or comb fiber goats three to six weeks prior to kidding.
- Sort pregnant does from open does.
- Supplement with selenium, if necessary, prior to due date.
- Pen unbred does with a buck for clean-up breeding.

Kids

- Breed late kids after seven months of age or at 75 pounds body weight.
- Castrate any late bucklings to be raised for meat.
- Tattoo or tag late kids.
- Vaccinate late kids according to schedule.

Herd

- Check for lice and delouse if needed.
- Monitor body condition of all goats and adjust supplemental feeding as required.
- Prepare sales materials and distribute sales lists.
- Cull unsound or inferior animals.
- Stock kidding supplies.
- Trim feet as needed.
- Clean equipment such as clippers so they are ready for kidding and show seasons.

Pastures

- Ensure kidding pastures have adequate shelter for freshening does.
- Evaluate forage and pasture conditions.
- Select kidding pasture if range kidding.

SPRING

Bucks

- Assess bucks that are available for breeding.
- Begin looking for replacement bucks.
- Give vitamin E-selenium (Bo-Se) or copper supplements if needed for your area.

Does

- Clip udders on nonfiber goats at least one week prior to due date.
- Freeze heat-treated colostrum for emergency use and CAE prevention.

  Tape doe's teats one week before due date.

  Segregate known CAE-positive does from the herd.

  Remove kids from doe immediately after birth.

  Feed heat-treated colostrum, pasteurized milk, CAE-free milk, or milk replacer
- Shear or comb fiber goats three to six weeks prior to kidding.
- Supplement lactating does to maintain milk production, increasing feed gradually.
- Supplement nondairy goats for at least four weeks post-kidding.
- Worm does at freshening or one to two weeks following kidding.

Kids

- Check the doe's teats and start milk flow so the kid may nurse easily.
- Feed colostrum within the first couple of hours after birth.
- Identify kids.
- Keep records of each freshening.
- Neuter males that are not being raised for breeding.
- Disbud kids.
- Monitor kids raised on the dam to be sure each is getting enough to eat.
- Pen kids with individual does for dam-raising when possible.
- Send registrations for new animals. Some registries, such as ADGA, have lower fees for early registration.
- Start coccidia prevention or test feces by three weeks of age.
- Supplement weak or slow-growing dam-raised kids with bottle-feeding.

Herd

- Monitor internal parasites by testing stool samples.
- Treat for worms before cleaning the barn or moving the goats to new pasture when testing shows high levels of worms.
- Shear nonpregnant fiber animals.

Housing & Equipment

- Clean and bed kidding areas several days before the doe's due date (if not completed in fall).
- Renew trailer licensing and affix tabs.

- Clean trailer or hauling equipment for market and show season.
- Order fly-control products.
- Sort show supplies and tack to be sure everything is in order.

Pastures

- Check and repair fencing.
- Check outside pens before moving animals. Repair fence as required.

SUMMER

Bucks

- Assess bucks that are available for breeding.
- Continue looking for replacement bucks.
- Market excess bucks and cull those that don't sell.
- Give vitamin E-selenium (Bo-Se) in selenium-deficient areas in late summer.

Does

- Cull does that have reproductive problems or ill health.
- Discontinue supplemental feeding to does that are at least four weeks postpartum and aren't being milked for dairy.

Kids

- Evaluate the kid crop to decide which will be sold and which kept for replacement.

- Monitor internal parasites by testing stool samples.
- Continue coccidia prevention or start testing feces by three weeks of age.
- Vaccinate at four, eight, and twelve weeks of age or as recommended by veterinarian.
- Worm at six to eight weeks as necessary.
- Tattoo or tag kids at four to six weeks of age.

Herd

- Clip or bathe goats to remove dead skin.
- Check for external parasites and ringworm.
- Keep animals hydrated and cool with water and ventilation during hot-weather hauling.
- Maintain adequate shade for animals during hot weather.
- Show season:
  - Do not overfill the udder on does, which can cause mastitis.
  - Make feed changes gradually.
  - Double-check show supplies.
  - Send show entries on time.
- Test stool samples for internal parasites.
- Treat for worms if necessary.
- Trim feet as necessary.
- Watch for foot rot in wet conditions.
- Make sure clean water is available at all times. Water usage increases due to lactation and hot weather.

Housing

- Continue fly-control measures.
- Keep pens well-bedded and clean frequently to prevent illness.

Pastures

- Rotate pastures every several weeks, if possible.

- Repair fencing and pens before moving animals.
- Watch pastures for poisonous plants and remove if present.

EQUIPMENT AND SUPPLY DEALERS

The dealers below have an internet presence and/or a catalog commonly used by goat owners. Mail order and the internet can offer greater convenience and possibly even lower prices than retail stores. Many of these suppliers began their business after their goat hobby got out of hand. It is interesting to visit these pages for the extra goat-raising information published alongside their product listings. In some cases, attached blogs allow customers to interact and share their knowledge. This list is by no means comprehensive. Some of these supplies can also be purchased through a larger marketplace such as Amazon or eBay.

Caprine Supply

A catalog and online company that specializes in all things goat. Founded in 1978, Caprine Supply is one of the first goat supply companies I used after I picked up a free catalog at one of my first goat shows. Supplies are now available for a variety of different livestock species, such as llamas, cows, and sheep, as well as goats. Their catalog and website also have a wealth of goat management information.
800-646-7736
www.caprinesupply.com

> BRING ME A BOWL OF COFFEE BEFORE I TURN INTO A GOAT.
> —JOHANN SEBASTIAN BACH

Hamby Dairy Supply

Located in Missouri, Hamby Dairy Supply specializes in milking supplies for goats, as well as sheep and cows. They carry goat milking machines plus Surge and NuPulse equipment. Their total supply catalog contains more than 3,000 items for everything from the small farmstead to the large commercial dairy. The Hamby family has been dairying since 1954 and began selling dairy equipment in 1991.
861-449-1314
www.hambydairysupply.com

Hoegger Farmyard

Formerly called Hoegger Supply Company, this long-running company published its first catalog of only thirty items in 1934. In 2011, they renamed their website Hoegger Farmyard. The company used to have an outstanding reputation and carried a wide assortment of cheesemaking, soapmaking, and goat-raising products, and their website has a treasure of articles, blogs, and information. Unfortunately, there have been reports the past few years of poor service and lack of communication with

customers. Several key family members years have gotten ill or passed away. I hope they can turn the company around; however, I must recommend you check recent reviews before using them for critical supplies.
770-703-3072
www.hoeggerfarmyard.com

Jeffers

Founded in 1975 by Dr. Keith Jeffers, PhD, in his basement, Jeffers has grown into one of the largest catalog and online sources for animal health products. As an advisor to local livestock owners, Dr. Jeffers saw that his farm clients had a need for convenient, cost-effective supplies. Starting out door-to-door, Jeffers grew a catalog business that eventually had three catalogs—*Equine Supplies, Pet Supplies,* and *Livestock Supplies*—all of which are available today. First online in 1999, they now have a large internet presence including an integrated eCommerce business combining all three catalogs, Facebook pages for each catalog, and lots of articles and tips in their blog. The company offers reasonable prices and good shipping. The company also has one brick-and-mortar store that is open seven days a week.
310 Saunders Rd. W.
Dothan, AL 36301
800-533-3377
www.jefferspet.com

Khimaira Farm

Khimaira is ancient Greek for "she-goat." More than forty years ago, Linda Campbell started raising goats. From there, she expanded, and she has the most amazing goat-related grouping of sites. Products and services range from supplies to books to art—plus everything in between. Khimaira Webservices supplies web hosting for farms of all sizes. The farm itself is now a popular wedding venue. Their catalog is my source for Weck Canners when they are available. I highly recommend visiting the Khimaira Farm main website and exploring the wide variety of offerings. Besides the businesses mentioned above, there are many goat resources—farms, organizations, knowledge databases, auctions, and links to almost anything related to goats. The Goat Connection site is especially good for information.
2974 Stonyman Rd.
Luray, VA 22835
540-743-4628
www.khimairafarm.com

KV Supply

KV Supply began in 1978 as a reseller of pet pharmaceuticals, vaccines, and health-related animal products. The company was sold in 2016 but continues to offer the same services and products. If your veterinarian will write a prescription for the drugs you need, this company will process and fill prescription medication.
844-493-6817
www.kvsupply.com

Lehman's

Lehman's was a small hardware store that opened in 1955 to serve the local Amish community in Kidron, Ohio. Today, Lehman's has a fascinating catalog with both a mail-order and internet presence. They are a good source for small dairy equipment, cheesemaking supplies, and non-electric appliances and sell lots of Made-in-USA items and old-fashioned farm tools.
800-438-5346
www.lehmans.com

Nasco Farm & Ranch

Another well-established agricultural supply house, Nasco was started in 1941 by a vocational agriculture teacher. Nasco now publishes more than thirty-five different catalogs. The livestock supply portion of the business is very complete, including milking machines and supplies, showing and fitting products, ranch and farm supplies, and educational materials.
800-558-9595
www.enasco.com/farmandranch

New England Cheesemaking Supply Company

Bob and Ricki Carroll started selling a few cheesemaking supplies in 1978 with a small catalog they sold for a quarter. The business

grew over the years and now has a online catalog with recipes and tips to go along with their books and products. For people who want hands-on experience, they also hold classes in cheesemaking near their Massachusetts business.
413-397-2012
www.cheesemaking.com

Northwest Pack Goats & Supplies

Since 1995, Northwest Pack Goat and Supplies has been making and improving pack goat equipment. Their website also includes links and information related to goat packing. The owners design and test their own equipment. Their offerings included saddle pads, saddles, feedbags, and even a medicine kit designed specifically for goats.
888-722-5462
www.northwestpackgoats.com

PBS Animal Health

An extensive offering of pet and livestock supplies and health products since 1941. PBS has a prescription drug section as well as over-the-counter medications. The company also has five stores in Ohio.
800-321-0235
www.pbsanimalhealth.com

Pet Supplies Delivered (formerly Omaha Vaccine Company)

This online store carries vaccines, dewormers, supplements, antibiotics, and pharmaceuticals for pets and horses. Forty years ago, their catalog was published as *Omaha Vaccine*. They have an impressive array of reasonably priced medical supplies that work for goats as well as for their specific customers, and also carry prescription pharmacy products.
11143 Mockingbird Dr.
Omaha, NE 68137
800-367-4444
www.petsuppliesdelivered.com

Premier1Supplies (Also Pipestone Veterinary Supply)

I used to recommend Premier1 for their exceptional array of electric fencing supplies. The founder of Premier1 worked in England before returning to the United States in 1977. When he and his wife couldn't find some of the equipment he had used in England, they imported products he missed for their Iowa farm. Starting out by ordering products for friends and neighbors led to a fence supply catalog, which leads us to their current expanding product line including fencing, animal identification, goats, sheep, and poultry.

In May 2018, the store portion of Pipestone Veterinary Supply was added to the Premier1 line. Now, in addition to the ElectroNet® fence, there is an expanded shopping area for goat health and care. I find the Farm to Table section of their catalog, which has dairy supplies (including cream separators), particularly fascinating.

Premier1 is partnered with Pipestone Veterinary Services and gives advice from experts. They offer "Ask a Sheep Expert," which lets owners ask questions about problems they're having with goats as well as sheep. Fencing, grazing, feeding, lambing, health, genetics, or general care questions are answered.
800-282-6631
www.premier1supplies.com

Valley Vet Supply

Founded in 1985, this veterinarian-owned-and-operated retailer has livestock and pet products. It carries everything from the common goat supplies and medicinals, as well as equipment from collars to folding milk stands. It also has a prescription medicine section.
800-419-9524
www.valleyvet.com

GLOSSARY

GOATS CAN'T TALK. THAT'S CRAZY!

—BRIAN FELLOWS, *SATURDAY NIGHT LIVE*

HOW TO TALK GOAT

ABOMASUM. In a ruminant's stomach, the fourth, or true, digestive compartment, which contains gastric juices and enzymes.

ABORTION. An abnormal or early termination of pregnancy.

ABSCESS. An enclosed collection of pus found in tissues or organs. Usually a sign of infection, although some are sterile.

ACCREDITED HERD. A herd of goats that has been annually tested for a specific disease (usually tuberculosis) and has been found free of the disease.

ACIDOSIS. An abnormal condition in which the rumen becomes too acidic, usually due to overcompensation of grain or rich forage or sudden dietary changes.

AFLATOXIN. A toxin produced by the molds *Aspergillus flavus* and *Aspergillus parasiticus* that can contaminate animal feed and cause serious problems.

AFTERBIRTH. The placenta and fetal membranes normally expelled from the doe's uterus within three to six hours of kidding.

ANEMIA. A deficiency of red blood cells or hemoglobin in the blood, often a sign in goats of parasites and a need for deworming.

ANTIBIOTIC. A medicine designed to kill or inhibit the growth of harmful microorganisms. It can be administered topically, orally, or by injection.

ANTIHELMENTHIC. A medicine used to remove or kill worms and other internal parasites.

ANTITOXIN. An antitoxin serum used to prevent or treat diseases caused by biological toxins, such as tetanus.

ARTIFICIAL INSEMINATION (AI). A breeding technique that involves depositing buck semen into the doe without sexual contact. This technique allows for increased genetic diversity.

BANDING. Neutering a buckling by the use of an elastrator castration band at the base of the testicles.

BILLY. An informal term for an adult male goat, typically an older nonwether in a meat or fiber herd. Dairy producers consider it a negative term.

BLEAT. The cry of a goat.

BLOAT. An acute indigestion characterized by swelling of the stomach due to excess gas from overeating new feeds or fresh forage.

BODY CONDITION SCORING (BCS). An evaluation system used to estimate the physical condition of a goat, typically on a scale of 1 to 5 (thin to fat).

BOLUS. An antibiotic administered orally as a large pill or as an intravenous injection.

BOTS. A parasitic disease caused by infestation of the stomach or intestines with botfly larvae, which crawl in through the nasal passages.

BROOD DOE. A doe bred for the purpose of continuing a desirable bloodline and genetics in her offspring.

BROWSE. Tender young grasses, leaves, twigs, and other vegetation favored by goats.

BRUCELLOSIS *(ALSO UNDULANT FEVER, MALTA FEVER, OR MEDITERRANEAN FEVER).* A disease caused by infection with bacteria of the *Brucella* group, frequently causing abortions in does and fever in humans.

BUCK. An adult male goat.

BUCKLING. An immature male goat.

BUTTERFAT. The fat content, or cream, of milk, used to make butter.

BUTT. To strike with the head or horns. Butting is the preferred method among goats for settling dominance issues.

CABRITO. Tender goat meat from a three-month-old milk-fed kid weighing 20 to 35 pounds, popular in Mexican dishes.

CAPE. Unshorn strip of hair left along a fiber goat's spine, about 6 to 8 inches wide, to protect the animal from becoming chilled after shearing.

CAPRICULTURE. Goat husbandry, including all aspects of raising and breeding goats.

CAPRINE. Of, relating to, or characteristic of a goat.

CAPRINE ARTHRITIS ENCEPHALITIS (CAE). Disease caused by a goat-specific lentovirus like HIV in humans. It causes chronic arthritis and sometimes progressive pneumonia or chronic mastitis in adult goats and sometimes a form or encephalitis in newborn kids. CAE prevention techniques include hand-rearing to prevent transmission of the disease from dam to kid.

CARRIER. (Genetics) An animal that carries a recessive gene and can produce offspring with a genetic defect. (Medicine) An animal that shows no symptoms of a disease but harbors infectious organisms that can infect other animals.

CASEOUS LYMPHADENITIS (CL). A highly contagious disease characterized by abscesses affecting the lymph nodes of sheep and goats. Caused by *Corynebacterium pseudotuberculosis.*

CASHMERE. A luxury fabric made of the fine underwool of non-Angora goats.

CERTIFIED HERD. A herd that has undergone an annual test for brucellosis and been found free of this disease.

CHEVON. Goat meat from a six- to nine-month-old kid weighing 50 to 60 pounds.

CHIVO. Goat meat from an older goat.

CHLAMYDIA. A microorganism that causes pneumonia, abortion, diarrhea, conjunctivitis, arthritis, and encephalitis in goats. There are many strains of this bacterium, which multiplies only in living cells and is spread by direct contact with fresh body secretions.

CHLAMYDIOSIS. An infectious disease that causes abortion in does. Pregnant women should not handle aborted caprine materials.

CLOSTRIDIA. Anaerobic bacteria against which most goats are vaccinated. Commonly found in the environment, this genus of bacteria includes the organisms responsible for tetanus and enterotoxemia.

COCCI. Bacteria that live in contaminated manure and cause coccidiosis in goats and other livestock. Some species (*Toxoplasma* and *Cryptosporidium*) are infectious to humans and other mammals; other types require a specific animal host and are not a problem for humans.

COCCIDIOSIS. A parasitic disease that destroys the lining of the small intestine, causing watery diarrhea, dehydration, and sometimes death. It is spread by ingestion of feces.

COLOSTRUM. The first milk produced by the doe after kidding; full of antibodies and minerals.

CORPUS LUTEUM. A gland that develops in the ovary following ovulation. It secretes the hormone progesterone and helps regulate the doe's estrous cycle.

CORTICOSTEROID (*ALSO CORTICOID*). A steroid—such as aldosterone, hydrocortisone, or cortisone—occurring in nature as a product of the adrenal cortex. Corticoids are given to goats as medication to relieve pain or inflammation.

CROSSBREEDING. Mating a buck and doe of different breeds to produce a hybrid offspring.

CRYPTOSPORIDIOSIS (*ALSO CRYPTO*). A parasitic organism that proliferates in the small intestine and causes diarrhea in goats and humans.

CUD. Food that is subjected to bacterial action in the rumen and then regurgitated to the mouth for more chewing. Cud chewing is unique in ruminants.

CULL. To remove from the herd goats that are unsound, physically inferior, below average in production, or fail breeder's specifications.

DAM. A mother goat.

DENTAL PAD. The hard gum pad on the upper jaw that substitutes for top front teeth in goats.

DISBUD. To prevent the horns on a kid from growing by cauterizing the horn buds with a hot iron.

DRENCH. To give a goat liquid medication by pouring into the mouth.

DRY. No longer yielding milk.

DRYLOT. A penned area for holding a herd for an extended period.

DUAL-PURPOSE. A breed that serves two purposes, such as milk and meat.

ELASTRATOR. A tool used to apply heavy rubber bands to the scrotum of a kid for castration.

ENCEPHALITIS. Inflammation of the brain, usually indicated in goats with severe signs such as fever, lack of coordination, and convulsions. Several diseases cause encephalitis, including polioencephalomalacia and listeriosis.

ENTERITIS. Inflammation of the intestinal tract.

ESCUTCHEON. Arched area at the back of the udder below the perineum where the hair grows up and out instead of down.

ESTRUS (*ALSO HEAT*). The breeding period preceding ovulation and during which the doe is receptive to mating the buck, usually lasting 24 to 36 hours. The peak of estrus, when the doe is most receptive to breeding, is called "standing heat."

EXTENDED LACTATION (*ALSO MILKING THROUGH*). The practice of milking a doe for more than one season without rebreeding.

FLEECE. The wool shorn from a fiber goat at one time.

FLUSHING. Management practice of improving a doe's nutrition just before ovulation to increase the number of eggs produced.

FORAGE. Food for goats that contains fiber, such as silage, hay, and pasture.

FRESHEN. To begin to produce milk. A first-time dam is called a "first freshener."

GASTROENTERITIS. Inflammation of the stomach and intestines.

GRADE. A goat produced by crossbreeding purebred stock with nonpedigreed stock. To upgrade or grade up is the sequential use of purebred animals over a series of generations to end with almost pure offspring.

HELMINTH. A parasitic worm.

HYBRID VIGOR. Increased vigor, meat production, longevity, reproduction, or other superior qualities arising from crossbreeding different goat breeds.

IODINE. An antiseptic for wounds.

JOHNE'S DISEASE. A chronic wasting disease of ruminants that causes diarrhea.

KED. A bloodsucking tick that pierces the skin, causing damage to a goatskin pelt.

KEMP. Short, hairy fibers found in mohair that have a hollow core and do not accept dye.

KIDDING. Bearing young, as in *the doe was kidding for the first time.*

LACTATION. The period during which a doe produces milk. In dairy goats, 305 days is the standard length of lactation.

LANDRACE. A race of animals developed in the wild through natural selection and ideally suited to the environment in which they live.

LIVER FLUKE. A parasitic leaf-shaped worm that rolls up like a scroll and infests the bile ducts or liver.

LUNGWORMS. Parasitic roundworms that infest the respiratory tract and lung tissue.

MANGE. A chronic skin disease caused by mites that infest and damage the skin and hair.

MANURE. The dung of livestock, commonly used to fertilize soil. Also known as stool, droppings, waste, excrement, fecal matter, poop, and nanny berries.

MASTITIS. Inflammation of the udder caused by bacterial infection.

METRITIS. Inflammation of the uterus.

MILKER. A goat that produces milk, most often used in reference to a dairy doe.

MILK FEVER (*ALSO PARTURIENT PARESIS*). A disease that affects dairy goats just after giving birth and at the start of lactation. A substantial drop in the blood calcium level interferes with nerve transmission, causing partial or almost total paralysis.

MILK REPLACER. An artificial milk substitute fed to kids when the dam's milk is unavailable or as part of a CAE prevention program.

MOHAIR. The fine hair of the Angora goat, which can be made into a luxury fabric; also, the fabric itself.

NANNY. An informal term for a female goat, typically a dam. Dairy producers consider it a negative term.

NITRATE POISONING. A condition caused by eating toxic level of nitrates from plants containing an excess of this product.

NOSE BOTS. Tiny larvae that crawl into the nasal passages.

OMASUM. The third compartment of a ruminant's stomach.

OPEN DOE. A female goat that has not been bred or has not become pregnant after breeding.

OVERSHOT JAW (*ALSO PARROT MOUTH*). A congential defect in which the upper jaw projects beyond the lower jaw.

PARALUMBAR FOSSA. Soft, hollow area high on either side of the goat below the loin; literal translation is "the depression next to the lumbar vertebrae." Can be palpated to detect rumen activity.

PARASITE. An organism that lives on or in another organism. External parasites, such as fleas, infest the skin, hair, and nasal and ear passages. Internal parasites, such as worms, infect the stomach, lungs, and intestines.

PARTURITION. The act of giving birth.

PINKEYE. Inflammation of the eye, most often caused in goats by mycoplasma or chlamydia. Often highly contagious between goats and transmissible to humans.

POLLED. Naturally hornless. A polled goat has two bumps where horns would typically grow.

PRECOCIOUS MILKER. A doe that comes into milk without being bred.

PREGNANCY TOXEMIA. A metabolic condition of pregnant does in which the blood contains toxins, generally caused by an energy-deficient diet during late pregnancy.

PROBIOTIC. A living organism, such as yeast, used to manipulate fermentation in the rumen.

PUREBRED. A goat of unmixed lineage.

RADDLE MARKER. A color crayon in a harness or colored paste put on a buck while pasture breeding.

RETICULUM. The second compartment of a ruminant's stomach, lined with a membrane having honeycombed ridges that increase the surface for absorption.

ROUGHAGE. Coarse, bulky plant matter that is high in fiber, such as hay and silage.

RUMEN. In a ruminant's stomach, the large first compartment, which contains microbes that break down forage and roughage.

RUMINANT. An animal that chews cud and has a four-compartment stomach. Goats, sheep, and cattle are ruminants.

SCOURS. Diarrhea in livestock, usually severe and watery.

SCURS. Incomplete horn growth resulting from inadequately removed horns.

SHIPPING FEVER. A respiratory disease usually acquired by a goat during transport.

SILAGE. Fodder prepared by storing and fermenting green forage plants in a silo or an airtight bag.

SIRE. A father goat.

SORE MOUTH (*ALSO CONTAGIOUS ECTHYMA OR ORF*). A highly contagious viral infection that causes scabs around the mouth, nostrils, and eyes and may affect the udders of lactating does.

STANCHION (*ALSO HEAD GATE*). A device used to secure goats in a stall or at a trough for feeding, milking, or work such as hoof trimming or artificial insemination.

STRIP. To remove the last drops of milk from the udder.

UNDERSHOT JAW. A congenital defect in which the lower jaw projects beyond the upper jaw and the lower teeth extend past the upper dental pad.

URINARY CALCULI. Stones of mineral salts in the urinary tract. Commonly found in wethers, urinary calculi are caused primarily by an imbalance of dietary calcium and phosphorus.

VACCINATION. An injection of a weakened or killed pathogen, used to stimulate antibody protection against the pathogen and produce immunity to a disease.

VAGINAL PROLAPSE. Protrusion of the vagina in does in late pregnancy.

WATTLES. Hair-covered appendages of flesh hanging from the neck of some goats. Wattles serve no real function; they are thought to have formerly served as scent glands, but this has not been proven.

WETHER. A castrated male goat.

WHITE MUSCLE DISEASE. A disease caused by a deficiency of selenium, vitamin E, or both, characterized by the degeneration of skeletal and cardiac muscles.

YEARLING. A goat that is between six and fifteen months of age.

YOLK. Wax or grease in mohair fleece. Yolk protects the fiber from sun and chemical damage.

INDEX

R

S

T

U

V

W

ABOUT THE AUTHOR

CAROL A. AMUNDSON has been raising goats since 1989. Her articles have appeared in *Goat Magazine* and *United Caprine News*. The former editor of the Minnesota Dairy Goat Association newsletter, *Gopher Goat Gossip*, Carol served two terms on the MDGA Board. Her initial goal to have "a few chickens, ducks, two goats, and maybe a calf and a sheep or two" turned into almost thirty years of goat husbandry experience. Those years involved building a grade A goat dairy, running a successful show string, and working at Poplar Hill, a 500-milker goat dairy. At times, Terrapin Acres, her 20-acre farm in Scandia, Minnesota, had over one hundred goats. Carol also worked as a clinical laboratory scientist. She currently spends most of her time on the farm where she and Dave raise cattle, pigs, a few chickens—and, of course, goats.